DESIGN IS THE PROBLEM
THE FUTURE OF DESIGN MUST BE SUSTAINABLE

Nathan Shedroff

Rosenfeld Media
Brooklyn, New York

Design Is the Problem: The Future of Design Must Be Sustainable
By Nathan Shedroff

Rosenfeld Media, LLC
705 Carroll Street, #2L
Brooklyn, New York
11215 USA

On the Web: www.rosenfeldmedia.com
Please send errors to: errata@rosenfeldmedia.com

Publisher: Louis Rosenfeld
Editor/Production Editor: Marta Justak
Interior Layout: Susan Honeywell
Cover Design: The Heads of State
Indexer: Nancy Guenther
Proofreader: Kezia Endsley

THIS WEEK: CH 10+11

DUE OCT 31:
    CH 12, 13, 14
    OCT 31: CH 15 + 19

Who Should Read This Book?

I believe that *design* is how we change the world. Designers are incredibly optimistic people—believing that they can, absolutely, change the world for the better. However, even without a design background or education, it doesn't mean you don't or can't make change in the world—and change for the better.

This book was written for the designer in all of us. While it is primarily targeted at those who call themselves "designers," it doesn't use any design jargon and is just as helpful to engineers, managers, students, and anyone who wants to build a better, more sustainable world. Whether you are involved with the creation of products, services, online experiences, events, environments, or mechanisms that drive systems (like the economy), understanding sustainability issues, frameworks, and strategies can help you create better solutions.

This book won't make you an expert—only experience and time can do that. However, it's designed to help you get acquainted quickly with state-of-the-art methods in the sustainability domains and to put you far out in front of most of your peers.

What's in This Book?

This book is a summary of what I feel are the most important approaches and aspects of sustainable design. It doesn't cover everything and it only goes into so much depth. However, it should serve not only as a good introduction, but also help people put sustainable design practices into their work, no matter what they do. It is also filled with a number of references and resources that will serve as great next steps as readers decide to explore more.

This book is organized into five main sections: Reduce, Reuse, Recycle, Restore, and Process.

These sections contain 15 chapters filled with advice and examples for making better strategies regarding sustainable solutions.

Section 1: Reduce

This section, "Reduce," focuses on the strategies for reducing material and energy impacts. It is the first place to start in designing or redesigning anything because reducing these impacts is imperative.

Section 2: Reuse

The two chapters in this section focus on strategies for making solutions (products, services, environments, or mechanisms) last longer and finding other uses when finished.

Section 3: Recycle

Just because something is recyclable, that doesn't mean it's actually recycled. There are several strategies for developing products to be more easily recycled that both reclaim as much residual value as possible and prevent virgin materials from needlessly being used.

Section 4: Restore

Developing a more sustainable product or service is important, but it's often not enough. Aside from reducing the impact our activities have on the future, we have a lot of work left in order to correct for the impacts we've had in the past. This section describes how we need to rethink systems in order to have positive results, rather than merely reducing our negative results.

Section 5: Process

Once we have an understanding of the strategies we can use, we need to understand how to put them into the processes we utilize every day. These three chapters describe how easily sustainability can be inserted into processes we already use, how to measure our results, and how to talk to other people about them.

What Comes with This Book?

This book's companion Web site (**rosenfeldmedia.com/books/ sustainable-design/**) contains pointers to useful sustainable design resources that I've found and written. It includes a calendar of my upcoming talks and a place for you to engage in discussion with others who are interested in sustainable design. We expect to post information on new sustainable design-related resources and special discounts for related applications as they become available. You can keep up with the site by subscribing to its RSS feed (**feeds.rosenfeldmedia.com/ sustainable-design/**).

We've also made the book's diagrams, screenshots, and other illustrations available under a Creative Commons license for you to download and use in your own presentations. Unfortunately, these don't include certain images we've only received permission to print for copyright reasons (such as the photos of cars or other examples). You'll find the book's original illustrations and diagrams in Flickr (**www.flickr. com/photos/rosenfeldmedia**).

Why a Book and Not Just a PDF?

Any book about sustainability runs the risk of not practicing what it preaches. It's a fair question to ask why this information is in a book at all—why it requires paper, printing, binding, and transporting to be effective. Why can't it be distributed and consumed electronically? The answer is that it could. In fact, it is. This book is available as a downloadable PDF file from **rosenfeldmedia.com**.

Why then a book as well? The answer is simple. In order to be effective, information must fit the needs of its audience. Paper is still so satisfying to most people—and especially designers—who prefer to purchase and read a book. Often, it's more tangible, and the material seems more real.

In addition, many readers still value physical products more than virtual ones. We still expect virtual products to be free (or mostly so) and physical products to cost something. Although the value of the information contained in both is the same, we value each medium differently in both economic and emotional terms. With the actual printed book, you can dog-ear it, write notes, and highlight passages. However, you can store the digital edition easily on your hard drive and keep it with you everywhere you go, use search and backup features, and share it (but please don't go crazy or my publisher won't ask me to write many more books). People can grab a book off a bookshelf, instead of looking for the PDF on a backup drive two years from now.

The sales of each edition will give us valuable data as to how willing people are to embrace fully-digital publishing and how viable it might be to publish without paper in the future. Perhaps our preferences will shift and books will primarily be electronic fairly soon. This would certainly improve their environmental impact.

FREQUENTLY ASKED QUESTIONS

What is sustainability?

Sustainability is an approach to design and development that focuses on environmental, social, and financial factors that are often never addressed. Sustainable solutions strive to improve the many systems that support our lives, including efficiently using capital and markets, effectively using natural resources, and reducing waste and toxins in the environment while not harming people in societies across the Earth. Sustainability focuses on efficient and effective solutions that are better for society, the environment, and companies. Sustainable organizations are often more successful when they pay attention to the details of waste and impacts, allowing them to function more cleanly, increase profit margins, and differentiate themselves from other organizations. See page xxi.

Why does being sustainable cost more—or does it?

Sustainable solutions don't always cost more than unsustainable ones. Many solutions are focused on energy and material efficiency, and these actually cost less up-front. Because our economic system rarely includes all of the social and environmental costs and impacts of products and services of the items we buy, the producers of sustainable solutions try to compensate for these costs. Doing this can cost more up-front, but often costs less over time since these solutions may prevent problems later. See pages 129 and 139.

Is climate change proven?

There is overwhelming evidence of climate change, leaving no doubt that the climate isn't what it used to be. What's at issue is whether this change is due to human activity or cyclic conditions in the environment. While there is no unequivocal proof that all of the changes are due to human activity, there is massive overlap between the evidence of climate change and the details of human activity. The majority of reputable scientists believes that an overwhelming amount of climate change is due to human activity, despite the lack of conclusive proof. For more information, see www. climatecrisis.net.

What's a carbon footprint?

One of the most important aspects of climate change seems to be the amount of carbon dioxide (CO_2) accumulating in the upper atmosphere. Scientific models explain why this may have an impact on climate change and how serious this is to the environment and our way of life on Earth. One of the most prominent strategies for reversing these effects is to reduce the amount of carbon dioxide we send into the environment. A carbon footprint is a way of estimating the amount of carbon dioxide our activities generate, and by understanding this, we can find ways to lower these emissions. It represents the total amount of carbon dioxide our activities generate—from heating our homes to driving cars to eating and drinking to working and living.

Carbon footprints are difficult to calculate exactly because there are so many variables. However, most carbon footprint calculators do a great job of estimating our personal or corporate carbon emissions by using averages. A great place to start estimating your carbon footprint is the calculator at Al Gore's Web site: **www.climatecrisis.net**.

Are hybrids really better than other cars?

Hybrid cars are certainly not a long-term answer. Hybrids are better than hydrogen cars or really big cars (like SUVs), and buying a hybrid car sends a powerful message to the automobile industry, as well as other companies and government agencies. However, in the long run, smaller gasoline cars are better for the environment overall, and electric cars are probably the best. See page 73.

Is nuclear power a more sustainable energy option?

Proponents of nuclear energy point to the reduced carbon dioxide emissions of generating electricity via nuclear power over traditional methods. But there are many other issues that need to be taken into account, including the amount of CO_2 generated in the mining, transportation, and refining of uranium, construction of the plants themselves, dealing with the waste over thousands of years, and the abysmal safety record of the nuclear industry with regard to workers, miners, and the environment. These additional costs make nuclear power a much weaker investment than spending less money on efficiency technologies and alternative energies, such as solar, wind, wave, hydro, and other renewable sources. See page 28.

What can I do to become more sustainable?

Because sustainability encompasses so many issues across the social, environmental, financial, and political spectra, there are many things each of us can do to quickly build a more efficient, effective, and sustainable world. We can start by learning about the issues and then evaluating our impacts with carbon footprint calculators (which is quick and easy to do). Next, we can simply make better choices, starting small, by changing our behaviors to be more sustainable. One of the most important things we can do is simply to be more efficient, using fewer materials and energy in our activities. This might mean wasting less food, not driving when it isn't necessary, turning lights and electronic equipment off when they're not being used, insulating our homes to be more efficient, and so on. When we purchase new things, we can look for more efficient versions or ones with higher ratings. Buying locally-produced items is generally more sustainable and helps build resilient local communities. Most of these changes don't even impact our quality of life much, and most sustainable solutions help us do more with less rather than just give us less overall. For more information, see www.climatecrisis.net/takeaction and www.wecansolveit.org/content/action.

As a designer, what can I do to make the world more sustainable?

First, designers can understand the breadth of sustainability and the strategies for developing more sustainable solutions. This is pretty easy (and is covered in this book). Next, designers can start using these strategies in their work, even if only a few at a time. We need to become advocates of sustainability issues for our own organizations and our clients, partners, and other stakeholders. We can address sustainability issues in our projects whether our clients and organizations appreciate them or not, making more sustainable solutions even when those around us don't do so.

Over time, designers can address more issues and integrate more strategies into their work naturally. This is easiest when all team members are aware of the issues and strategies and when sustainability becomes part of the process. Ultimately, sustainability is most powerful when it becomes part of an organization's values and mission, but we don't need to wait for this to begin in order to have an impact now. See page 266.

TABLE OF CONTENTS

CHAPTER 3
What Are the Approaches to Sustainability?

Reduce

CHAPTER 4
Design for Use

CHAPTER 5
Dematerialization

Process

FOREWORD

Design.

The word conjures images of effete eccentrics imposing cuboidal-built environments, clashing color, tortured fashion, and over rated celebrity upon the jaded palates of urbanites with too much money.

This book is not about that.

Nathan brings the competence of a mechanic, the mind of an engineer, the training of an MBA, and the pen of a poet to a topic long abandoned to people with delusions of adequacy. He talks about solutions, ones that deliver desired outcomes, and how to implement them.

His focus is the design of a world that works; as he says, "Don't do things today that make tomorrow worse." Good advice.

This book presents a systems approach to crafting answers to the really big challenges, including how to meet human needs on a planet on which all major ecosystems are in decline, and it's a race to see which will melt first, the Arctic or the economy.

Most of us, if we think about design at all, consider color, or perhaps shape. But reflect that every human artifact was designed by someone. This person made deliberate choices about the utility of the object, the materials used to make it, the manufacturing process chosen, the length of its useful life, and what would happen to it after it was no longer needed. Consciously or by choosing to ignore opportunities, we have created a world in which half a trillion tons of stuff is pulled from the Earth each year, put through various resource-crunching activities, shaped (at great energy cost) into a form, and then thrown away. Of all this stuff, less than one percent is still in use six months after sale. All the rest is waste.

At the moment of conception of an idea, a design, a thought of a product or a process, 80 to 90 percent of the lifetime cost of that widget, program, or pickup truck was committed.

Investing in how designers think, in how we all approach a new idea, is thus crucial if life as we know it is to thrive on this planet.

Nathan has given us the mental model to begin that exploration. He does so with a soft touch, but a ruthless honesty. One of my favorites of his lines is, "Get over the guilt or shock or outrage or embarrassment or disagreement now, because none of it will be useful. We have a lot of work to do."

It is almost axiomatic that designers are arrogant and indulgent. Nathan is not. He delivers an outstanding primer on the precepts of sustainability, the challenges facing the world, and pragmatic answers in a playful and accessible manner. This book should be part of any curriculum on design, innovation, business, environmental studies, marketing, public policy, engineering, organizational development, and the now rapidly emergent field of sustainability. It should be on the desk of CEOs of all companies that make or deliver anything. It will be required reading for all of my students, and a frequently recommended treat for the companies with whom I consult.

It should be the next book you buy.

—L. Hunter Lovins
    Author of *Natural Capitalism and Sustainability*
    Chair, Presidio School of Management

Introduction

This isn't a book about sustainable design. Instead, it's a book about how the design industry can approach the world in a more sustainable way. Design is interconnected—to engineering, management, production, customer experiences, and to the planet. Discussing and comprehending the relationship between design and sustainability requires a systems perspective to see these relationships clearly.

I hate discussions that start with definitions, but the truth is that the terms "sustainable" and "design" at the beginning of the 21st century are both malleable and subjective enough to warrant an explanation. However, I'll try to get the definitions out of the way quickly and efficiently to get to the larger discussion.

> **This ... is a book about how the design industry can approach the world in a more sustainable way.**

What Is Sustainability?

Design is in great transition, thankfully. Traditionally, design has been practiced with a focus on appearance, whether it is represented in graphic, interior, industrial, fashion, furniture, automotive, marine, or any other kind of design. In truth, design has never been merely about appearance, although that's been the most prominent phenomenon throughout its history. In addition, other disciplines use the word "design" to describe other functions, such as structuring databases, systems, services, or organizations (further confusing its use and meaning). But there have been moments in design's past where truly great designers showed us that design was also concerned with performance, understanding, communication, emotion, desire, meaning, and humanity itself, even though these haven't been the most lasting movements.

Ultimately, this is the design that I want to speak about in this book—design that encompasses the synthesis of usefulness, usability, desirability, appropriateness, balance, and systems that lead to better solutions, more opportunities, and better conditions, no matter what the endeavor or domain.

In the end, there is no reason that great design can't be beautiful *and* meaningful and sustainable.

Sustainability, too, isn't well defined—even by its own practitioners. To many, it is synonymous with *green*[1] (not that *green* is any more clear) or *eco*, meaning *the environment*. To others, it connotes bleeding-heart nouveau hippies, who seem more concerned with plants and animals than people. Sometimes, it's portrayed as a way to promote old, flawed economics as a way of ensuring "business as usual." Often, it's a threat to a way of life that can only, possibly, mean less of everything. Or it can be interpreted as a rational blend of constraints both large and small and a way to serve human needs on all levels, as well as those of other systems.

Sustainability means more than all of this. It refers to human and financial issues as much as environmental ones. The systems perspective inherent in sustainability encompasses cultural impacts as well as ecological ones, financial constraints as well as physical limits, and heritage and legacy as well as a perspective about the future.

The most agreed-upon definition of sustainability comes from the Brundtland Commission[2] and dates back to 1987:

(Use and) development that meets the needs of the present without compromising the ability of future generations to meet their own needs.

---

[1] The term *green* has become so problematic that Adam Werbach, CEO of Act Now Productions, suggests using *blue* instead. This alludes to a natural color that is devoid of hippie overtones, friendly to business, and is ubiquitous on the planet (the sky). www.saatchis.com/birthofblue

[2] For more definitions of sustainability terms, consult the *Dictionary of Sustainable Management*. www.sustainabilitydictionary.com

Put simply: *Don't do things today that make tomorrow worse.*

There, that doesn't sound so silly, or dumb, or dangerous, does it? It sounds like common sense. Unfortunately, designers have been very bad about this. The fact that engineers and politicians and marketers and accountants and business leaders and educators and everyone else have been equally bad doesn't absolve us from this reality—or our responsibility.

An even deeper meaning to sustainability points to the need to restore natural, social, and economic systems (and the effect they've had on society, nature, and markets), and not merely "fix" them to make them perform better. This concept of restoration will be addressed later in this book, but first, let's be sure we understand how to fix the systems themselves to reduce the damage created and to stop it from advancing.

The essence of this definition, which may not be obvious immediately, is that needs aren't just human, they're systemic. Even if you only care *about* humans, in order to care *for* humans, you need to take care of the system—(the environment) that you live in. And this environment doesn't include just the closed system we call the planet Earth. It also includes the human systems we live in— our societies—and the forming, changing, and constantly evolving values, ethics, religion, and culture that encompass these societies. We aren't separable from each other, and we can't ignore the effects of the whole—nor should we. Indeed, that's where much of the humor, cleverness, and fun lie. To take a systems perspective acknowledges that individual perspectives don't necessarily speak for or represent the whole when talking about the environment, the economy or markets, or any aspect of society. Yet, to take systemic action requires that we act in concert with others, despite our differing approaches. This is what makes sustainability difficult. It is also what poses the biggest design opportunity.

Sustainability, then, needs to address people (known collectively as "human capital"), our cultures, our needs and desires, and the environment that sustains us (known as "natural capital"), as well as the financial mechanisms (known as "financial capital") that make most forms of design thrive. Solutions that don't encompass or work in concert with others across these aspects of our lives significantly reduce their ability to succeed.

Therefore, designers need to find ways to address all of these issues in their solutions.

Part of the Problem or Part of the Solution?

Despite how optimistic, idealistic, and future-oriented most designers are, design has sometimes created big problems in the world. Even where our best intentions have been engaged, our outcomes have often fallen short—sometimes making matters worse—because we didn't see the whole picture when creating what we envisioned. Where our best intentions haven't been engaged or where we haven't been well-enough informed, design has been dismal. The same is often true of other business functions like marketing, sales, operations, engineering, and so on. We are often responsible for making people feel terrible about themselves, only redeemable by buying this product or that service. In addition, we too often contribute to a philosophy of "more is more" when it doesn't deliver more value and when it simply wastes resources of all kinds.

Designers are taught to make "new" when it isn't really better or when "old" doesn't need replacing. Often, designers are complacent when their engineering and marketing colleagues suggest (or insist on) low quality over longevity, cheap materials, or bad usability.

A sad truth is that almost every solution designed today, even the most "sustainable" one, has more of a negative impact on the planet than a positive one. This means that the world would be better off if most of what was designed was never produced. This is changing, and it doesn't have to be the case in the future, but we have a long way to go in order to change this pattern.

> **Designers are taught to make "new" when it isn't really better or when "old" doesn't need replacing.**

For example, Fritjof Capra's definition of sustainability is "human activity that does not interfere with nature's inherent ability to sustain life." This isn't a bad definition, but it's hazy as to how it assesses what does and doesn't impede the environment. Can you name one thing produced that could be said *not* to "impede the Earth's ability to support life"? If you take a systems perspective and acknowledge that all things are connected through the system, then every product that humans have created, from the industrial revolution on, could be cast as impeding the natural environment's ability to function properly. This is why I don't consider this to be a helpful guideline for design. That said, it punctuates the impact designers have on the world and why we so drastically need to reframe the solutions we create into a larger context.

All design disciplines have too often focused on creating meaningless, disposable, trend-laden fashion items. My own design education was in automobile design, a discipline that's never been a flag-bearer of function over form. The very term *planned obsolescence* (for which design and marketing should be forever remorseful for inventing and promoting) came from the car industry. But graphic design is no better, nor fashion design, nor even interaction design. We're all guilty of having our collective attention diverted too often by trends, operational difficulties, or financial challenges. This means that all designers, no matter the experience or domain, can make things better. We can all be *part of the solution* as the popular saying from the 60s goes.

My friend Eric once explained to me that "fashion is for ugly people to have something special about them." He's right on more levels than he realizes.

For starters, it's no secret that the most beautiful thing you can wear is an authentic smile, and that the most beautiful people among us would be just as beautiful barefoot or in burlap, rather than in Manolo Blahniks. We're a weird species that will spend fortunes to have the *latest* things, only to throw them away a season later, or spend our time and money on things that cover our bodies rather than make our bodies as healthy, fit, and beautiful as they naturally could be.

There are no ugly people—only impatient or mean or intolerant ones. This is the truth of fashion and design.

Quick, what's the most beautiful complexion, the best height, the correct size of nose, and the right waistline? There is no more an "ugly" than there is one way to be in the world. People and industries create and maintain the whole concept of ugly just so they can sell often ridiculous, temporary products and services to insecure, frustrated, scared, and vulnerable people—and none more so than tweens and teens. This is where the most damage is done and, in the interest of more sustainable futures, where we need to start correcting the problem.

I have no problem with fashion or the part of design that focuses on appearance. Trends, in fact, can be fun, like a party or a film or a destination you visit. But let's not mistake them for more substantial design. They aren't a replacement for quality, valuable, or meaningful solutions. Fashion, at its best, is about responding to people's desires, aspirations, and the reality of materials and the human body within a cultural context in ways that accentuate our better selves. Too often, design has been the mechanism of "cheap and dirty" or "fast and dirty," and it has been used as a weapon to hurt people (emotionally and even physically) just as much as it has been used to enable and inspire them.

There are no ugly people—only impatient or mean or intolerant ones. This is the truth of fashion and design.

Makeup and Stilettos

Consider the mainstays of women's fashions. Seasons and trends come and go but two constants sold to women as the foundation of beauty are makeup and stiletto heels. For centuries, women have been told that they are at their most beautiful when they cover their skin with a patina of pore-clogging, often toxic, substances. Ironically and sadly, these very materials often ruin their skin years before the natural effects of age occur, making for worse complexions and requiring (seemingly) even more of the stuff that caused the problem in the first place. Foundation and heavy makeup has, thankfully, fallen out of fashion in the past few decades, yielding to a desire for more "natural" beauty, but many women are still taught that the first step to looking pretty is to cover up their skin and start with a blank canvas. Yet, when we think of authentic beauty, the examples mostly offered are those of youthful and athletic people. Advertisements would have you believe that the most beautiful people in the world rely on makeup, diets, and questionable home exercise equipment when these same models themselves don't use what they promote.

Stilettos and most other "high" heels are the same. They exist to fool women that they will be more attractive by buying products that actually ruin their feet, posture, and backs over time. While they sometimes serve functions other than fashion (such as giving women a bit more height), this could be accomplished by fashionably thick-soled shoes instead. That women regularly torture and destroy their bodies in the name of fashion, not to mention often flirting with accidents from tripping, speaks to how deeply this false view of beauty is rooted in our culture.

Sure, it can be fun to "dress up" in costumes once in a while, using makeup and clothing to temporarily play a role. I have no issue with this. However, those who habitually trade their health and natural beauty in for fashion solutions that actually harm both are playing a sad game. And those who promote this behavior are fooling themselves if they think they're making people more beautiful.

Design at its best, however, focuses on people and seeks to understand what it can offer them to make their lives better in some way. Despite the celebrity designers foisted upon us by the design industry, successful design isn't about some personal vision spun out and overlaid onto the world to make it seem shiny and new. Successful design is careful and considered. It responds to customers/users/participants/people, market, company, brand, environment, channel, culture, materials, and context. The most successful design is inseparable from these criteria. The most meaningful design is culturally and personally relevant, and we respond to it on the deepest levels. The best design also has a future. It is sustainable.

Design can be all of this. It *needs* to be all of this.

It should be clear by now what I mean by *design is the problem*. Design that is about appearance, or margins, or offerings and market segments, and not about real people—their needs, abilities, desires, emotions, and so on—that's the design that is the problem. The design that is about systems solutions, intent, appropriate and knowledgeable integration of people, planet, and profit, and the design that, above all, cares about customers as people and not merely consumers—that's the design that can lead to healthy, sustainable solutions.

Get over the guilt or shock or outrage or embarrassment or disagreement now, because none of it will be useful to you going forward. And we have a lot of work to do.

Design that is only about appearance, or margins, or offerings and market segments, and not about real people—their needs, abilities, desires, emotions, and so on.—that's the design that is the problem.

CHAPTER 1

What Is Sustainability?

1

Because sustainability means different things to different groups and individuals, this makes it sometimes difficult to discuss it, so I'll just give you my definition. For example, I met a "conservative environmentalist" once who was chiefly focused on climate and carbon issues. His advice to me on my own project, Reveal (a sustainability rating and labeling system for consumers—see Chapter 17), was to measure only climate-related issues (like carbon dioxide production) and leave the social issues for people to figure out on their own. My response was that if customers discovered that the product they believed to be "better" due to the ratings actually had questionable social impacts (such as animal cruelty or child or slave labor), all trust would dissipate for both those products and the rating system.

Therefore, the only way to approach sustainability effectively is from a systems perspective. We need to consider a wide perspective before diving into details. Because most things are connected to most other things, to design anything effectively requires considering what it connects to. This necessitates folding financial, social, and environmental issues together—at least at some point. For sure, these issues are widely different and require specialized knowledge and solutions, but no solution can be addressed effectively without considering its impact across all three areas. You will find, in practice, that you can't solve everything. However, you will need to be ready to address why you can't do everything. And, just possibly, in the process you may find that you can address more aspects than everyone around you suspected—which is quite often the result of good design. Constraints are a challenge for designers, not a limitation.

One serious problem for designers is that, even with a systems approach, there are few tools in existence that wrap these issues together. Instead, designers must learn to patch together a series of disparate approaches, understandings, and frameworks in order to build a complete solution. The good news is that these different frameworks are compatible, as you'll see in Chapter 3, "What Are the Approaches to Sustainability?". Their vocabulary may be slightly different, but a meta-framework can be built that organizes a coherent and somewhat complete approach to sustainability. This is what I've tried to do in this book.

To address the three domains of human, natural, and financial capital, it's important to understand some of the issues central to each. Different contexts can change the priorities or urgencies of each, but all of these themes are relevant and important to understanding why sustainability is imperative.

> **You will find, in practice, that you can't solve everything. However, you will need to be ready to address why you can't do everything. And, just possibly, in the process you may find that you can address more aspects than everyone around you suspected—which is quite often the result of good design.**

Although this is the most important part of the story, it's still often not good enough for many clients, companies, or business people who just don't understand why these issues should be dealt with by themselves and their businesses. For them, there is another set of issues that might be more influential in changing their opinions. I won't go into these in much detail since they aren't really design issues but following is a list for you to explore further. The better you understand these issues, the more persuasively you can help your company's and client's goals align with those of sustainability.

- Even the business press regularly reports on sustainability and social responsibility issues. This not only creates validation for sustainability efforts, but it can also exert important peer pressure on business leaders and managers to "get on the bandwagon" before their competitors.

- At its heart, sustainability is about efficiency. There's not a manager or leader in the world who isn't looking for his or her organization to be more efficient. Being able to reduce expenses—especially at a time when costs of almost everything are skyrocketing—is always good business.

- Sustainability is not a fad (changes that are fleeting), it's a trend (changes that will endure). Whether in the consumer markets or the business markets, more and more customers are concerned with environmental and social issues. There are opportunities for businesses to take advantage of this trend by aligning their brand and promotional efforts with those of the society as a whole—as long as it's authentic. Otherwise, being exposed for greenwashing is probably worse than not doing anything at all (more about this in Chapter 18).

- Customers, shareholders, investors, and other stakeholders are quickly joining the sustainability wave—no matter how they articulate it. Investors (in particular pension, retirement, and other large funds) are moving their money to investments that align with their longer-term values.

- For many companies, sustainability focuses on risk mitigation (reducing financial and other costs associated with potential risks), especially for insurance, reinsurance, and health-related companies. Any company with a stake in the long term is looking to mitigate its operational and financial risks over that amount of time.

- Governments at all levels are also quickly shifting regulations— especially at the state and local levels—to reinforce and reward organizations that take sustainability seriously. The business world can fight this all they want, but if they lose, they'll have to make the changes anyway—after their competitors have already mounted the learning curve ahead of them.

What Is a Systems Perspective?

Because sustainability requires a systems perspective, sustainable design must also address the system, whether it is a market, an ecosystem, a social system, or the entire world. This allows the design process of sustainability to address the environment, markets, companies, and people.

It's easy to understand the concept of a systems perspective: the system is the sum total of everything affected by an activity. A systems perspective requires an appreciation (at a minimum) and an understanding (at best) of how various systems interact with each other. These include environmental, financial, and social systems.

It can be more difficult to understand the boundaries of the systems affected by a particular action and within a particular domain. In addition, measuring these interactions is often faulty, at best. And keeping this perspective in mind during the entire development process isn't easy. However, it's possible. You don't have to create perfect solutions in every way. What you have to do is make conscious, informed considerations across the spectrum of financial, environmental, and social issues, just like you should across the dimensions of customer experience. Your solutions should be better than the baseline (what others devise without considering these issues) and can improve over time.

Consider the impact that population has on everything (see Figure 1.1). All issues addressed being equal, the biggest one is probably overpopulation.

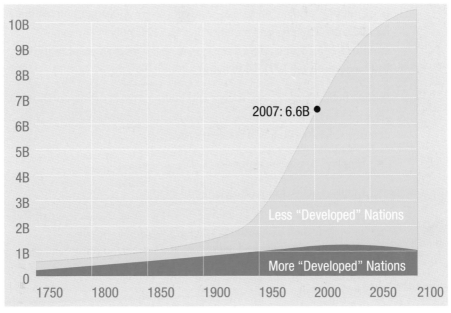

FIGURE 1.1
World population.

Solving this one challenge would have the greatest impact on all of the others. If the world had even one-half of the population it presently has, many of our more pressing problems would simply disappear. At half the population, food and resources would probably be abundant. Population, however, takes literally generations to change. If people, collectively, gave birth to only half the number of children now, it would be 40 or 50 years before the effects would truly be visible. And how can we drastically reduce the earth's population in any socially acceptable way? China instituted a one-child-per-couple policy to great outcry, and its effect has yet to be truly felt. Could that solution be successful in other cultures? What are our other options?

Diversity and Resiliency

Perhaps the best way to judge any system or solution is to assess how resilient it is. Systems that are resilient have a greater chance of lasting, evolving, and responding to change. Nature has evolved and proven to be tremendously resilient, which has allowed it to grow, change, and continue for millennia. Cultures and societies that are resilient have the same attributes. Moreover, each of these systems has a mechanism of resiliency and longevity (such as DNA in nature).

In their book, *Brittle Power*, Amory and Hunter Lovins define systems that are *resilient* as *able to withstand large disturbances from the outside.* They describe several attributes of resilient systems:

- Employs passive behavior

- Employs active feedback (learns and adapts)

- Detects faults early

- High substitutability and redundancy

- Optional interconnectedness (can connect or disconnect when needed)

- Promotes diversity

- Makes use of standardization

- Dispersed

- Hierarchically embedded

- Stable and flexible

- Simple

- Makes limited demands on social stability

- Accessible

Corporations are more resilient organizations since they've been granted personhood, limited liability, and immortality (though these can cause other problems). Since corporations can survive their founders, as well as shield their owners from the actions of their managers, they attract more diverse capital and can govern themselves in ways that offer a great deal of flexibility and strategies.

The problem with diversity is that it's often not valued in society.

Diversity is one strategy for resiliency since it allows multiple solutions and approaches to solve or respond to the same challenge. A diverse community is more able to weather a storm, a bad growing season, a financial hardship, or a cultural crisis. Companies with diverse resources have more tools to respond to market challenges, customer whims, or corporate missions.

The problem with diversity is that it's often not valued in society. Consider how many varieties of apples (~7500) or rice (~80,000) have evolved. Before humans, these varieties helped all of nature (plants, animals, bacteria, fungi, monera, and so on) respond to often hazardous changes in the environment due to weather, tectonic events, or those from space. Today, however, human societies have reduced the number of varieties of almost everything natural to just a few. The vast majority of the U.S. agriculture system, for example, is built on five to six varieties of apples, three to four types of cherries, and fewer than 20 types of grains—often without even preserving discarded varieties in case we need them in the future. It may seem like good business from a financial perspective (fewer options to choose between), but it's very bad business if you want a resilient marketplace. Your design solutions and processes will have an impact on resiliency and diversity, and you should consider if that impact is positive or negative.

Resiliency is often missed when evaluating our markets, environments, or systems. Because it's abstract, we focus on lower-level issues that contribute to overall resiliency, but often don't address the challenges we face. For example, in trying to solve poverty, we often focus on the symptoms and not the cause. In our fervor, we rush to simple solutions that are standardized because they are easier. We also look for quick fixes to satisfy the urgency we feel. But poverty is not an easy problem to solve. To be effective, poverty needs diverse solutions across the entire system (and often with a long time to take hold). If you want your time and energy to be truly effective, your design solutions will also need to reflect how they contribute to sustainable, lasting, systemic change—not merely the most visible symptoms.

Community resiliency is achieved through a wide spectrum of issues: financial stability and opportunity, diverse education, reliable safety and security, values embedded in systems and markets, and so on. We can't, for example, ever expect to be safe from terrorism simply by installing alarms and cameras everywhere. Terrorism isn't the product of anything that can be watched by a camera or triggered by an alarm. Yet, these easy solutions are the ones we jump at, and they provide temporary emotional peace—until they fail. *Periodically, you should evaluate whether you're truly working on a solution at the right level to make the change intended.* Often, our projects are defined in terms and at levels that are more easily grasped by organizational functions but aren't pushing at the pressure points that are needed to make change. Your system perspective is what will guide you to understanding the network of interactions that challenge your customers and clients to determine whether the solution is even capable of affecting the kind of change desired.

Periodically, you should evaluate whether you're truly working on a solution at the right level to make the change intended.

Centralization and Decentralization

While it's often easier to manage a few, centralized systems, these are often less sustainable solutions because, though strong, when they fail, the rest of the system fails with them. This is why a tree falling on a power line in Washington state can trigger a power outage over most of the Western United States. Centralization was the management approach that governed society and culture throughout much of human history (certainly Western history) and was responsible for creating the Industrial Age when it was applied to production. Everything from central banking to centralized power plants to large corporations with central management to centralized distribution systems to centralized education to the "hub and spoke" air transportation model reflects the thinking that centralization is best. And, from a purely management perspective, it often is. But just because it has been popular doesn't make it the best approach.

Centralization is not without serious faults. <u>Centralized decision-making often doesn't reflect local expertise, knowledge, or understanding.</u> Centralized distribution, combined with the standardization necessary with economies of scale often reduces choice, favoring quantity over variety. Centralized power (such as a coal-powered electricity plant) often produces power more efficiently that must be transmitted over greater distances (which reduces efficiency) and can reduce pollution in some communities but instead concentrate it in others. In addition, centralized power is often less resilient since fewer, larger power plants are vastly more vulnerable to accidents, outages, and attacks than many, smaller ones distributed throughout the service area.

It is because of the negative impacts on diversity and resiliency that centralization is often less sustainable. Decentralized systems for everything from manufacturing to distribution to energy generation to political rule tend to be more sustainable. Consider how unresponsive centralized government often is for local issues. Or consider how much more resilient a community's power grid would be if it had a mix of energy inputs (especially if these were renewable) spread across a power network, available locally where it is used. Natural gas turbines, geothermal and hydrothermal generation, co-generation (creating energy from waste), solar, wind, and so on can all exist easily within most communities without adverse health risks or other community concerns. Where possible, generating power where it is used has always been an efficient solution (mills have been situated next to rivers that could provide water power for centuries). However, organizations (whether corporations, NGOs, or governments) that thrive because of their centralized control are often the most vocal opponents to decentralized solutions because their advantage is threatened, despite the beneficial aspects these solutions may have for everyone else.

It is because of the negative impacts on diversity and resiliency that centralization is often less sustainable. Decentralized systems for everything from manufacturing to distribution to energy generation to political rule tend to be more sustainable.

To be sure, decentralization itself also has problems. Chief among them are standardization and communication. While decentralization can increase resiliency (and often equitable opportunity), it requires standards and increased communication in order to function. The benefits, however, are often increased efficiency in management, more resilient operation in failure, and more innovated techniques in solutions generated. For example, it's often easy for local communities to establish their own standards that may not be consistent, fair, or interoperable in a larger context. This was largely the case with every developing technology, from screw sizes to electricity current to railroad track gauges to education standards to laws to money. Decentralized solutions are often problematic at their onset (this is especially the case with new technologies) until standards are established. Progress is often retarded as competing solutions compete on low-level features and performance, that is until standards are established cooperatively or competitively. For example, consider software file formats. Until a standard was established for interchanging files and communicating with other equipment, say PostScript, applications couldn't universally talk to printers, typesetters, or other equipment. It wasn't practical to even work on advanced applications like page layout applications, image editing applications, or content management systems until these standards were established, despite the fact that they were envisioned long before they were able to be implemented.

Designers need to be aware of how their solutions inhibit or reinforce centralization—and be ready to defend why and whether their solutions are improvements and for whom these improvements benefit.

Cooperation and Competition

Competition, while a powerful motivator in innovation, is not the only ingredient needed for successful, sophisticated solutions. Despite how we characterize innovation and design, nothing is created in a vacuum, and no solution is successful without cooperation between people, including design teams, partners, supply chains, and customers.

> Competition, while a powerful motivator in innovation, is not the only ingredient needed for successful, sophisticated solutions.

Cooperation is often misunderstood as being unnatural. Competition has been drummed into our heads as the driver of natural evolution for so long that we often classify cooperation as merely a human invention. The fact, however, is that all sophisticated systems, including nature, have required cooperation on lower levels in order to support competition on new, higher levels. In time, often, these higher levels are standardized, and cooperation leads to new innovations that compete at yet even higher levels.

For example, this book would not be possible without a high degree of cooperation in production, manufacturing, distribution, and even language. If we didn't agree on grammar, syntax, vocabulary, and morphology as much as we do, we wouldn't be able to communicate with language. Language, in fact, wouldn't exist, and without it we wouldn't be able to discuss such abstract concepts as cooperation versus competition. This is always the case when we learn a new language or when we learn about new systems, like sustainability.

Cooperation is often misunderstood as being unnatural.

Nature is awash with examples of cooperation (evolution itself wouldn't be possible without it), and it is the only reason why complexity develops. Competition may be the mechanism by which new innovations succeed or fail, but cooperation is the foundation on which innovations occur. In our drive to support innovation and improvement (evolution, in other words), we often discount cooperation, which directly limits our ability to create complexity. We shouldn't be afraid of cooperating on standards, systems, and understandings because this is a necessary precursor to higher-order development and advancement. For all design—especially sustainable design—it's imperative that we cooperate on some levels in order to succeed at others. As designers, we must be aware of where we (and our clients) need to cooperate in order to understand the best opportunities for competition.

Competition may be the mechanism by which new innovations succeed or fail, but cooperation is the foundation on which innovations occur.

In addition, while cooperation means *working together, collaboration* implies *working together toward a common goal.* As designers, we both cooperate and collaborate with a range of *stakeholders.* Not everyone has to be working toward the same goals in order to cooperate. However, you will find many stakeholders whose goals already align if made clear: customers, suppliers, retailers, waste collectors, and so on. But you won't find these opportunities to cooperate and collaborate if you don't look for them.

Ecological Vitality

There is no question that an unhealthy, unstable environment decreases efficiency and our ability to create stable, healthy societies and communities. Yet, human history is filled with examples where we do just the opposite. For a variety of reasons, we have accepted the destruction of healthy, vital habitats for ourselves and the natural systems that we rely upon. This has to change. Increasing climate change—indeed, climate crisis—is requiring us to take a systems perspective in order to create healthy, more vital natural capital for which to support human life and activities. Some of these concerns include:

- Habitat destruction and collapse

- Topsoil depletion (which retards our ability to grow food)

- Habitat alteration

- Reduced biodiversity

- Climate change (also known as *global warming, global weirding,* and *climate crisis*)

- Ozone depletion

- Fresh water supplies

- Air pollution

- Toxic pollutions (including carcinogens, acid rain, and the by-products of industrial and agricultural chemicals)

- Over-concentration of substances (too much of even good materials, in too high a concentration, or in the wrong places, are just as toxic as harmful materials)

- Resource depletion (such as oil or water)

- Destruction of eco-services (such as the environment's ability to clean air and water, and shade us from harmful ultraviolet rays)

For all of these issues that affect the environment, they all directly affect human health as much as they affect the health of plant, animal, and other life in nature.

The stress our activities have placed on the environment by our population has endangered not only specific species, but whole systems. Food, water, and energy are intimately interconnected, although our policies treat them as separate and unrelated. If you're interested in the more details on this topic, a great place to start is the U.N. Millennium Assessment at **www. millenniumassessment.org.**

Social Vitality

To be considered sustainable and just, many designers require products to have a positive impact on the society they are serving (as well as those who helped create them). Product planning must embrace the concept of stakeholder involvement and incorporate social responsibility. Most people in the West are appalled and embarrassed when they find that products they've purchased were made with child or slave labor. Certainly, these aren't values they promote in their own communities. However, most never bother to inquire whether these conditions exist for the goods they buy and merely wait for the media to inform them of what does or doesn't reflect their social values. Consumers often rely on assumptions that companies they trust—especially, large, known companies—wouldn't sell such goods, but this assumption is often a mistake.

For example, it's not acceptable to most people for a disposable diaper to be considered sustainable if it's dangerous or detrimental to the environment. However, it's also not acceptable, for many, if that product is dangerous to the people who use it. To still more, it can't be sustainable if those making it are at risk. Still others question whether the product itself truly fulfills a sustainable role in society.

... all communities and individuals have different social values.

To complicate this further, all communities and individuals have different social values. It's impossible for a company to offer solutions that satisfy everyone. So they often don't bother satisfying anything but the law. Smaller companies will sometimes specialize in offerings for customers with specific social values (such as religious restrictions), and our governments (at all levels) will often legislate certain standards that a majority can agree upon, but this is often not enough. We'll see in the next chapter the myriad social issues in this space, and you'll understand the difficulty organizations have trying to satisfy customers' social concerns.

Social concerns are issues for sustainable designers—and there are a lot of them. Some affect corporate policy—either for our own firms or our clients. Others operate at the product, service, or event level and govern the design of such solutions. All require careful cooperation with a variety of experts in other roles, including executives, engineers, marketers, and managers of all types (hiring, operations, finance, sales, etc.).

These are also issues of vision and mission for a company—whether a design firm, a client, or the design department within a larger corporation. This is where designers must learn to be strategic and communicate to business leaders in business language. Designers are often disempowered—frequently by their own doing. When designers fail to understand the issues, vocabulary, and concerns of business leaders, they're not equipped to participate in strategic discussions that decide the organization's mission, vision, goals, or offerings. They must be content dealing with the results of these decisions at the implementation (or tactical) level and design the best solution they can that fits the already-specified parameters. Instead, designers should seek to involve themselves in the strategic discussions that determine not only what the offerings' parameters are but what to offer the market in the first place. This is where designers can have the most impact. (More discussion of this is in Chapter 16, "Innovating Solutions.")

Financial Vitality

For sure, innovative solutions, no matter how sustainable, can't be effective if they aren't financially viable. While designers, traditionally, eschew these considerations or assume others are "on top of them," this further disempowers their work. Currently, the economy is stacked against sustainable solutions because it doesn't recognize or value the *true cost* (that which totals social and environmental costs in addition to financial costs) of the products and services that are created, deployed, and disposed of. This makes it even more difficult to design in a sustainable fashion. However, those designers who understand the mechanisms that support or challenge the products and services they develop, will find more opportunities to have innovative ways of solving challenges at all levels of sustainability (environmental and social as well).

> **Currently, the economy is stacked against sustainable solutions because it doesn't recognize or value the *true cost* (that which totals social and environmental costs in addition to financial costs) of the products and services that are created, deployed, and disposed of.**

As I said earlier in this chapter, at the heart of sustainability is efficiency, and this always makes economic sense. Designers who reduce the amount of energy or materials in their solutions are inherently making more sustainably—financially—viable results. In this way, we can often take on more challenges across the array of sustainable issues and still deliver financially attractive answers.

But financial viability means more than just a better bottom line. As we've seen recently in the global financial markets of fall 2008, lack of transparency and accountability can create whole financial systems that aren't real, sustainable, or viable. The basics of supply and demand, credits and debits, loans and savings weren't at issue here (although even these have their problems). Instead, a shadow world of unaccountable value was created and then allowed to grow unsustainably until the entire house of cards came crashing down. In the wake, the good investments and value suffers along with the bad. This, too, is a lack of sustainable design. Those who developed these mechanisms didn't have a systems perspective, lacked ethics and accountability, and developed a system that wasn't healthy for any system, let alone the financial system itself. This is what can happen when sustainability isn't part of the criteria in a design and development process.

> ... financial viability means more than just a better bottom line. As we've seen recently in the global financial markets, lack of transparency and accountability can create whole financial systems that aren't real, sustainable, or viable.

An Ecosystem of Stakeholders

While we've traditionally considered the important players in the development process to be the client (or company) and the customer, these are not the only two actors to consider. The design industry has recently started to recognize the importance of deep customer understanding (often called *design* or *user research*) in the development of successful solutions. However, from a systems perspective, there are other stakeholders to be aware of as well.

Traditional approaches to business define shareholders as the only population to consider when making business decisions. More enlightened approaches to business include other groups, such as employees and customers. The most effective organizations, however, have learned to consider input, needs, and cooperation with suppliers, distributors, retailers, and other business partners throughout the supply chain. You can see where this is going. The more systems-oriented you are and the more you consider the full spectrum of sustainable issues (managing and using human, natural, and financial capital), the wider the circle of concerns, issues, and actors to involve. These are all called *stakeholders* and can include, in addition to those named above, any of the following: creditors, communities, government courts and departments (city, state, federal, and international), banks, media, institutional investors and fund managers, labor unions, insurers and re-insurers, NGOs, media, business groups, trade associations, competitors, the general public, and the environment itself (local, regional, and global).

Stakeholders can have all sorts of impact and exert considerable influence in remarkable ways.

Of course, no company or designer can include all of these stakeholders in all of its decisions, nor would this be appropriate. But, often, a particular stakeholder group can exert unexpected and significant power on a company due to its behavior and offerings. (Think of citizen action groups thwarting plans for a new dump in their neighborhood.) Stakeholders can have all sorts of impact and exert considerable influence in remarkable ways. They can form coalitions with each other to increase their influence, like when one labor union joins with another or when an NGO teams up with a trade association. While most of these stakeholders don't have direct corporate influence (such as voting on corporate decisions), they can exercise their power via the courts, government lobbying, economic activities, and social activities, such as boycotts, protests, and creating awareness in the media and public.

Smart companies keep an eye on as many stakeholders and their concerns as possible and manage company activities and policy accordingly to reduce the need to respond to unexpected stakeholder involvement. Designers, too, need to be aware of the range of stakeholders and their concerns if they hope to create sustainable solutions that improve conditions across environmental, social, and financial challenges.

A Careful Balance

Lastly, no decision is perfect or comes without consequences. Design requires decisions that narrow possibilities, ultimately until there is one solution. Designing more sustainable offerings may require you to balance inputs and outcomes and, often, compromise. It's rare, in fact, that you'll achieve everything that you want.

For example, some recycled materials have lower strength, higher weight, or less perfect consistency. This may require your product to have thicker construction or lower tolerances. Parts may need to be larger or heavier. These outcomes may be the result of necessary dematerialization or sourcing materials from a source with better environmental or social behavior. Other factors may require you to choose between a longer or shorter product life, increased or decreased carbon footprint or recyclability, or less efficient energy use. Mobile products, especially, might cause you to choose more expensive, lighter, and more durable materials since these products need to be as small, light, and robust as possible, or they don't get purchased. But these may also limit longevity or recyclability. In addition, more sustainable materials may be more expensive than less sustainable ones, driving up the total product cost.

> **Designing more sustainable offerings may require you to balance inputs and outcomes and, often, compromise.**

You may not be able to educate your customers enough for them to appreciate or expect your new approach or solution to compete with those they already know and trust. For this reason alone, an optimized solution, from a sustainability perspective, may not be successful. Getting too far in front of customers or the market can be more disastrous than being too far behind (since it often results in disastrous sales and failure of products, services, and companies).

As I wrote in the "Introduction," there's no perfect solution. There are only choices balancing results that, hopefully, lead to a product, service, or experience with better performance of some kind.

"What is the use of a house if you haven't got a tolerable planet to put it on?"

—Henry David Thoreau in a letter to Harrison Blake (20 May 1860); published in Familiar Letters (1865)

CHAPTER 2

How Is Sustainability Measured?

You Get What You Measure

This is the reality of most of the world—and especially the business world. Measures are seductive. Attaching a score or value to something often makes it seem more legitimate, accurate, and valid. Even if the measurement or scoring system is hopelessly flawed or the things being measured are fundamentally so qualitative that they resist quantitative measurement—scores, ratings, and numbers make us more comfortable, and often influence decisions more than any other issue. If there is a number, for example, we can track its rise or fall and compare it to similar numbers for related solutions.

Unfortunately, this means that things that can't be measured easily because they're so abstract or too big to measure (like happiness or environmental impact) often get ignored in decision making. At first, we may try to keep in mind their value alongside more quantitative measures, but after a while we forget to include them, and they disappear from not only our consciousness, but also our development criteria and management procedures. Throughout history, many of the benefits that design can bring have been difficult to measure, especially the ones that are projected into the future. This inability to project has made it difficult for designers to convince all manner of people, from managers to business leaders, to investors, to educators, and even to government officials the value that specific solutions and the design process itself can add.

This, too, is what has happened with social and environmental issues in business. There have been recent attempts to measure the financial value of social and environmental issues, and these are important steps. For example, if you can show the financial savings in energy efficiency for a skyscraper, especially as compared to the nominal differences in construction costs, you've got a great justification for building a more "green" building. Likewise, if you can show that your publicity solution will likely cause one million people to replace two standard incandescent light bulbs with LED bulbs a year, saving electricity in a municipal power system and preventing tons of carbon and sulfur from being released into the atmosphere, then you've got a great case for social venture investment.

Social Measures

But how does one value social issues? How do you measure the financial benefit of saving a life or not causing pain to an animal? And should you? Even in purely financial terms (such as calculating the lifetime earning capacity for a Nigerian child saved from a disease), these approaches ignore the emotional, ethical, and meaning value of saving or improving a life. There have been no measures, to date, capable of tallying the social benefits of most social and environmental issues. In a 1997 article in *Nature* magazine, 13 experts from a variety of institutes and organizations calculated the total value of ecosystem services to be between $16 and $54 *trillion*. The average, which is on the conservative side, is $33T. This represents the value that nature provides to us (individuals, businesses, governments, and organizations alike).

> In a 1997 article in *Nature* magazine, 13 experts from a variety of institutes and organizations calculated the total value of ecosystem services to be between $16 and $54 trillion.

Ecosystem services include the following areas:

- Regulation of atmospheric chemical composition.

- Regulation of climate (including global to local temperature, precipitation, etc.)

- Regulation of climate disturbances (dampening fluctuations in storms, floods, draught, etc.).

- Regulation of water for storage and access (both agricultural and industrial uses).

- Control of erosion and sediment.

- Control of soil (formation and retention).

- Regulation of nutrient cycles (storage, cycling, processing, and acquisition).

- Treatment of waste (pollution control, detoxification, etc.).

- Control of biological systems—except humans (regulation of populations).

- Provision of habitats for migrating species.

- Production of food.

- Supply of raw materials (organic and inorganic).

- Supply of genetic resources.

- Recreational uses (including sports and tourism).

- Cultural uses (aesthetic, artistic, educational, spiritual, scientific, etc.).

Those numbers measuring the financial impact of environmental issues are already staggeringly persuasive in many cases, but perhaps we shouldn't try to measure social benefits in such a way.[1]

Just to give you a taste of what I'm talking about, let's just look at a list of the potential social issues that concern people (see Table 2.1). For sure, few people track all of these issues, but all of them show up on someone's radar screen, in one SRI (Socially Responsible Investment Fund) screen or another, or become the focus of a protest at some point.

---

1 The value of the world's ecosystem services and natural capital, *Nature*, Vol 387, May 199

TABLE 2.1: 2008 Qualifying and Rating Criteria from Highwater Research

GOOD COMPANY
Intention

ENVIRONMENTAL
Genetic Modification
Nuclear Power
Fossil Fuels
Clearcut Logging
Hazardous Waste
Industrial Farming
Animal Cruelty

SOCIAL
Human Rights
Unethical Conduct
Gambling
Tobacco
Weapons
Fast Food
Alcohol
Sexually Explicit Material
Explicit Violence

LEADERSHIP
Financial Management
Social and Environmental Commitment
Social and Environmental Execution
Management / Board Integrity
Stakeholder Engagement

COMMUNITY
Community Relations
Economic Impact
Philanthropy

CUSTOMERS
Customer Satisfaction
Customer Safety
Disclosure and Labeling

EMPLOYEES
Working Conditions
Employee Relations
Compensation and Benefits
Employee Wellness

PRODUCT AND SERVICES
Product Accessibility
Societal Contribution
Product Design
Extended Producer Responsibility

SUPPLY CHAIN
Supplier Standards and Selection
Supplier Chain Engagement
Supply Chain Transparency

WOMEN AND CHILDREN
Employment Practices
Women in Leadership

DIVERSITY
Employment Practices
Diverse Leadership

MATERIALS
Raw Material Demand
Material Waste
Molecular Waste

ENERGY
Energy Demand
Energy Sourcing
Energy Efficiency

WATER
Water Use
Water Quality

CLIMATE
Greenhouse Gases
Policy Impacts

Any one of these criteria can break out into detailed subcriteria. For example, Animal Cruelty might refer to voluntary or mandatory animal testing of medicines or products, owning zoos or circuses with animals, using animal products in the production of products and services, using animal labor, raising animals for food, cruel living conditions or abuse, promoting or reducing animal diversity, and so on. And none of these are standardized or commonly agreed-upon criteria within the categories. Different individuals and groups will expect different performance and adherence.

As you can see, there are *hundreds* of potential issues of interest to customers, investors, users, and other audiences. Designers don't need to have a deep knowledge of every issue, but they should take a pass over the list every now and then to be familiar with the variety of issues important to people who are impacted by their solutions. There is no way to address them all, because many are in opposition to one another. For example, some religious-based organizations and investment groups don't want to invest in pornography, while others see it as a litmus test of Freedom of Speech. Some don't want to support Lesbian/Gay/Transgender issues (or, more accurately, support companies that support these causes), while others expressly seek out those that do. Ultimately, you'll have to make your own choices as to what is important to you, your company, its brand, your clients, and your customers, but you need to have a broad awareness before you can do so.

Aside from the long list of social issues, the difficulty of creating a measurement system for them, and their often mutually-opposed imperatives, we need to address the culture of "optimization" that pervades such viewpoints. In personal finance, for example, the goal is to optimize money in absolute terms. Financial managers are required by law to help their clients realize the highest possible returns. This focus purely on financial measures ignores those same customers' social values. It's been very recent that investment funds target specific social issues so that people and organizations don't transgress their goals in the pursuit of simply maximizing their funds. For example, labor unions and pension funds are realizing that investing in companies responsible for accelerating the transfer of jobs overseas isn't in the long-term interest of their organizations. Despite potentially higher returns, many are reprioritizing where they invest in order not to contribute to the problems they face. The same was done by organizations and individuals in the 1980s who didn't want to support apartheid in South Africa.

More recently, personal financial advisors have helped their clients to balance quality of life issues over optimization of financial returns. In other words, many people are content to earn a little less if it makes them happier to support organizations, companies, products, and goals that are important to them rather than contributing to the very things that upset them in the world. Likewise, the culture of working too much in order to provide the most money possible for our families is changing in favor of earning less in order to spend more time with our families. These people can't measure their happiness or satisfaction in financial terms, but they're often happily accepting lower pay in favor of more meaningful lives.

Environmental Measures

Environmental criteria are usually both easier to measure and easier to address than social issues—if only because the emotion and ambiguity surrounding most social issues aren't part of the picture. Environmental issues are often measured and addressed in terms of materials and energy use (both amount and type of each). There is often little disagreement that toxins, for example, are bad for people, but there may be considerable disagreement over how they should be measured, how bad they may be for people, what levels are "acceptable," and how these concerns weigh against financial, sourcing, or access trade-offs.

Like social criteria, environmental measures can be exhaustive with little agreement over priorities or validation, and include the following areas to consider:

- Consumption and conservation of energy

- Consumption and conservation of water

- Consumption and conservation of air

- Consumption and conservation of organic materials (including food)

- Consumption and conservation of inorganic materials

- Consumption and conservation of recycled and upcycled materials

- Consumption, conservation, and source of energy (and percentage renewable)

- Production and reduction of pollution and other toxic emissions to air, water, and land (there are thousands of potential substances under this category alone)

- Production and reduction of waste

- Production and reduction of product packaging (including biodegradable)

- Consumption and conservation of transportation (including energy, source, amount, and emissions)

- Area of land disturbed, protected, or restored

- Disturbance or preservation of biodiversity

Example: Nuclear Power

For example, nuclear power is once again considered a prominent alternative, despite the disregard it was met with in the 1970s. This is because it's now being touted as a more environmentally beneficial solution since it emits far fewer greenhouse gases during electricity generation than coal or other traditional power plants. It is widely accepted as a somewhat dangerous, potentially problematic, but manageable source of generating electricity. Radiation isn't easily dealt with, especially in nuclear waste and maintenance materials, and expensive solutions are needed to contain, control, and shield both people and the environment from its harm. The dialogue about using nuclear power—and expanding it—centers on weighing these risks against the rewards, as well as the risks inherent in other forms of power generation. These are just some of the issues involved.

Example: Nuclear Power (continued)

PROS

- Lower carbon dioxide (and other greenhouse gases) released into the atmosphere in power generation.

- Low operating costs (relatively).

- Known, developed technology "ready" for market.

- Large power-generating capacity able to meet industrial and city needs (as opposed to low-power technologies like solar that might meet only local, residential, or office needs but cannot generate power for heavy manufacturing).

- Existing and future nuclear waste can be reduced through waste recycling and reprocessing, similar to Japan and the EU (at added cost).

CONS

- High construction costs due to complex radiation containment systems and procedures.

- High subsidies needed for construction and operation, as well as loan guarantees.

- Subsidies and investment could be spent on other solutions (such as renewable energy systems).

- High-known risks in an accident.

- Unknown risks.

- Long construction time.

- Target for terrorism (as are all centralized power generation sources).

- Waivers are required to limit liability of companies in the event of an accident. (This means that either no one will be responsible for physical, environmental, or health damages in the case of an accident or leakage over time from waste storage, or that the government will ultimately have to cover the cost of any damages.)

Example: Nuclear Power (continued)

- Nuclear is a centralized power source requiring large infrastructure, investment, and coordination where decentralized sources (including solar and wind) can be more efficient, less costly, and more resilient.

- Uranium sources are just as finite as other fuel sources, such as coal, natural gas, etc., and are expensive to mine, refine, and transport, and produce considerable environmental waste (including greenhouse gasses) during all of these processes.

- The majority of known uranium around the world lies under land controlled by tribes or indigenous peoples who don't support it being mined from the earth.

- The legacy of environmental contamination and health costs for miners and mines has been catastrophic.

- Waste lasts 200–500 thousand years.

- There are no operating long-term waste storage sites in the U.S. One is in development, but its capacity is already oversubscribed. Yucca Mountain is in danger of contaminating ground water to a large water basin, affecting millions of people. It's difficult, if not impossible, for the U.S. to impose its will on the state of Nevada (or other places) if they don't want to host long-term storage of waste.

- There are no operating "next generation" reactors, such as high-temperature breeder reactors and particle-beam activated reactors, that are reported to produce less waste and have reduced safety concerns. Even if these technologies were ready, they wouldn't be deployable commercially for another two decades.

- Shipping nuclear waste internationally poses an increased potential threat to interception to terrorism (though this has not happened yet with any of the waste shipped by other countries). Increasing the amount of waste shipped, particularly in less secure countries, is seen as a significant increase in risk to nuclear terrorism.

This is just a taste of the complexity of issues involved with nuclear power. Every issue, from a systems perspective, quickly becomes a complex discussion juxtaposing factors from financial, environmental, social, and political realms.

Financial Measures

How we measure financial returns says a lot about our values, just as it does with social and environmental returns. We can't ignore financial measures—even in nonprofit endeavors—but we have to be aware of what to measure and what not to measure. Sustainability asks us to consider a host of nonfinancial issues in designing and developing new solutions. These may not be able to be quantified in numbers or currency. Regardless, even when they can't be represented in financial terms, they are still important.

Currently, economic activities are measured in time-honored ways, such as assets, liabilities, expenses, profits, interest, and so on. These, however, are based on assumptions about how markets work that date back decades and ignore environmental and social values. For example, our markets assume that the value of money decreases over time. This is why interest exists: to compensate for getting back less value than what you loaned. However, this doesn't have to be the case. It's simply an agreement in our economic models. We could have just as easily created a model that assumed the opposite. Again, this is about design.

A common approach to business and finances in the West was famously described by Milton Friedman: "The business of business is business." This is an attempt to dismiss social and environmental concerns from business and the marketplace, partly because it's so difficult to measure and integrate these issues. This attitude has dismissed businesses from any requirement of responsibility based on social and environmental outcomes and their effects.

A common approach to business and finances in the West was famously described by Milton Friedman: "The business of business is business."

In my opinion, this is a failed frame, partly because it simply hasn't been the case that companies have behaved exemplarily in these aspects, and partly because social values have always been in the marketplace, just undercover or under special circumstances. Consider, for example, a U.S. company trying to argue that it should be able to sell key strategic products (computers, munitions, energy, materials, services, etc.) to an enemy during war. If Milton Friedman were right, no free-market capitalist should have a problem with this proposition. However, most businesspeople (conservative or not) would never agree to this.

As a result, the measures we use to assess financial progress are equally flawed in their assumptions. Economists call the deficiencies in their economic model *externalities*.[2] These externalities hide implausible assumptions such as "buyers make rational decisions" and "buyers make decisions based on perfect knowledge." Only recently has the field of economics begun to correct these fallacies that figure into not only our prevailing economic theory and model but also our most popular calculations and measures. As a result, our policy decisions reflect flawed assumptions and track inaccurate, and sometimes inappropriate, measures of success.

A good example of these popular measures is the Gross National Product (GNP). This measure is used as the chief indicator of a nation's economic health. However, profound impacts that are seen as deleterious to the social fabric and quality of life of nations (such as divorce, natural disasters, environmental degradation, etc.) actually contribute to a better GNP. This is because the total cost of these circumstances isn't calculated in anything but the narrowest financial terms. Instead, new (and probably still imperfect) measures, such as the Genuine Progress Indicator (GPI) attempt to compensate for the uncalculated social and environmental impacts. A comparison of the GPI and GNP for the last 50 years, for example, shows a significant difference in their measures (see Figure 2.1).

---

2 Externalities are basically a cheat. Real markets are incredibly complex—much more so than all economic models. When economics acknowledges the factors that influence real market action (such as the decision-making process customers go through or the social and environmental costs of deforestation) but that aren't included in economic models, they are referred to as externalities (since they're missing—or external—to the model). The history of economic models is that the most difficult (and often most important) factors governing economies have been externalities missing from the models used to make economic policy.

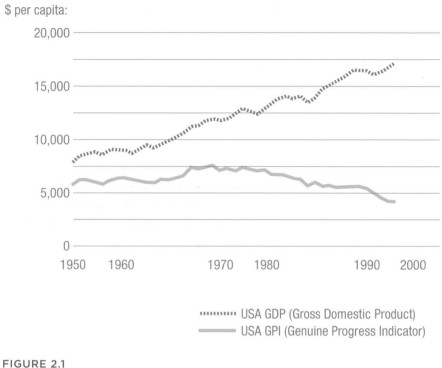

$ per capita:

20,000

15,000

10,000

5,000

0

1950   1960   1970   1980   1990   2000

••••••••• USA GDP (Gross Domestic Product)
━━━━━ USA GPI (Genuine Progress Indicator)

FIGURE 2.1
GNP vs. GPI.

Putting all of these measures together requires a new model for accounting. Where organizations only need to track financial measures currently, new instruments like the *triple bottom line* require organizations to track measurements across all three categories. But, this is still not enough. Simply tracking the numbers doesn't mean they will be consulted. The *integrated bottom line* attempts to combine these values so that environmental and social values aren't easily ignored. However, not only are there no standards for how to account for these other measures, but most companies don't even track the data necessary to do so—yet.

Integrated Bottom Line

In the late 80s, environmentalists started urging companies to use a "triple bottom line" to manage profit while also protecting people and the planet. Unfortunately, this approach simply bolted the promotion of environment and social issues onto standard balance sheets as cost centers, reducing the traditional measure of profit.

A more useful approach is the "integrated bottom line," a term coined by socially responsible investment expert, Theo Ferguson. This recognizes that profit is an important measure, but only one of many criteria that helps organizations provide enduring value. The Integrated Bottom Line approach integrates costs and benefits for social and environmental criteria into the traditional structure of the income statement and balance sheet, eliminating the separation of these issues and directly highlighting where the value of sustainable business impacts business performance and value:

- Raise profitability by cutting energy and materials costs in industrial processes.

- Drive innovation, leading to increased top-line revenues.

- Improve facilities design and management, and fleet management to increase effectiveness.

- Lead the effort and have first-mover advantage.

- Reduce risk and unbooked legal liabilities.

- Improve access to capital.

- Improve corporate governance.

- Enhance core business value through better government and stakeholder relations.

- Build brand equity by differentiating product and service offerings and enhancing reputation.

- Build competitive advantage through increased market share.

Integrated Bottom Line (continued)

- Increase a company's ability to attract and retain the best talent and detract from lost knowledge and expertise or increased training costs from replacement personnel.

- Increase employee productivity and health.

- Improve communication, creativity, and morale in the workplace.

- Build better supply chain and stakeholder relations.

Companies on the Dow Jones Sustainability Index have outperformed the general market by integrating the thinking behind the integrated bottom line across all aspects of their business. Recently, Goldman Sachs found that leading companies in environmental, social, and good governance policies outperformed the MSCI world index of stocks by 25 percent since 2005. Seventy-two percent of the companies on the list outperformed industry peers.[3]

---

3 Alderton, Margo, "Recent report finds corporations that lead in corporate responsibility also lead in the market," *Socially Responsible Investing.* 07-11 17:57, also at www.csrwire.com/companyprofile?id=4489.

Another approach is called *total cost assessment.* This calculation requires a team within an organization to generate and share data from all areas, including manufacturing, design, R&D, engineering, transportation, marketing, facilities, purchasing, external affairs, inventory, strategy and management (insurance, legal, accounting), and so on. In addition to current accounting of revenues, costs, employees, supply chains, etc., it proposes adding value criteria from market and competitors, governments, NGOs, and other stakeholders, as well as currently unaccounted values for things like product disposal (both costs and savings).

Other, new measures are being proposed (such as the Sultan of Bhutan's Gross National Happiness indicator), but we need more proposals. We've yet to establish satisfactory measures that calculate total social and environmental costs in relation to traditional monetary costs, and until we do, we won't have the tools necessary to make smart decisions about our futures.

Putting It All Together

One way to test these concepts is to consider how they interact with real questions people have in the world. We'll look quickly at two—not necessarily to answer them definitively but to uncover, in a real-world context, the complexities encountered when asking such seemingly simple and obvious questions.

Quick, which bag is better for the environment, paper or plastic (see Figure 2.2)? Are you sure?

FIGURE 2.2
Paper or plastic—the eternal question.

Because even the experts can't agree. The ubiquitous question asked of shoppers by grocery baggers has been turned around in the past few years as it pertains to environmental choices. These are exceedingly simple options, most often comprising one part and fairly simple manufacturing processes. Their function, too, is incredibly simple and obvious in almost all cases. This makes it an easy question to start with. Let's look at the pros and cons of each.

	Paper	Plastic
Pros	Made from a renewable resource (trees) Biodegradable	Much lighter than paper bag Despite being made from fossil fuels, uses considerably less material and releases significantly less greenhouse gas in the manufacturing and transportation
Cons	Trees are renewable only if replanted and carefully managed Biodegradable only if not put in a landfill (basically, nothing in a land fill ever degrades)	Made from fossil fuels

One camp contends that paper bags are better because they're biodegradable and don't rely solely on petroleum products for their manufacture. They also claim that paper bags don't pose as much of a danger to wildfire (which turns out not to be much of a problem, in actuality[4]). So paper bags must be better.

Which bag is better for the environment, *paper* or *plastic*?

However, the other side claims that plastic bags are better than paper because they weigh so much less per bag that the gasoline and diesel

---

4 There's actually little to no evidence that wind-strewn bags, though unsightly, have contributed to the destruction of habitats and resulted in the death of wild birds and animals. March 8, 2008 © Copyright Times Newspapers Ltd., **www.timesonline.co.uk/tol/news/environment/ article3508263.ece.**

burned to move them around (from the factory where they're made to the
stores where they're used to your car taking your groceries home) save
so much in emissions of carbon dioxide that they more than make up for
the oil used to make the plastic. In addition, the production of paper bags
generates 70 percent more air pollutants and 50 times more water pollutants
than does the production of plastic bags because four times as much energy
is needed to produce paper bags, and 85 times as much is needed to recycle
them. Okay, then plastic bags must be better.

Unfortunately, both sides are correct.

But how can this be? One has to be better than the other, right?

This is one of the problems with sustainability. The issues are so complex
and interconnected that even the experts are having difficulty coming to
conclusions. Customers simply want to know which is the better product to
buy. Most are, overwhelmingly, interested in buying products that support
their values. However, we can't give them the information they desire
because we don't yet know it ourselves.

There may be an even better answer, though. How about no bag? Or a reusable bag?

While some communities, such as the city of San Francisco, have begun
to ban plastic bags, there's no consensus among experts as to which is
better. The ones who care most about greenhouse gases contributing to
global climate change promote the use of plastic bags simply because
their weight is so much less than if we were to add up the miles traveled
from manufacturing to store to your car to your home, the resulting fuel
and emissions saved alone is drastic over paper bags. In contrast, those
promoting paper bags point to paper being biodegradable and sourced from
a renewable resource (trees). However, this is true only if the trees felled to
make the bags come from well-managed forests and are actually replaced
and cared for until adulthood. Likewise, biodegrading waste is preferable
to static waste, but people should know that almost nothing degrades in a

landfill since it isn't permeated by water, sunlight, and insects and bacteria that could break down the trash.

The answer isn't easy. In fact, it's probably a tie. If the paper bags are composted correctly and not put in a landfill, then they may be a better proposition—especially if they're reused as many times as possible. However, most bags aren't composted. So plastic bags, even though it seems less intuitive, may be the better option in the short run, especially as we move to reduce global levels of greenhouse gasses, like carbon dioxide.

There may be an even better answer, though. How about no bag? Or a reusable bag? Many times, stores automatically put purchases in bags, even when they're not needed (such as when there's only one item, or when customers already have another bag with them, or when the product itself comes in protective packaging—like an orange). Convenience is important but watching a grocery line for even a few minutes will show the amount of waste in bags that are used unnecessarily.

Dematerialization (see Chapter 5) teaches us that the less we use, the better. So, no bag is often the best answer. This highlights one of the first principles we can rely on for progress: less is often better. However, *less* is a tricky word. This principle isn't saying that we should do with less (functionality), but use less material to deliver the same—or even better— performance. That's the true meaning of *less is more*.

For a moment, consider another question: "Which is better for the environment, a paper cup or a ceramic mug?"

Finding the point at which one solution is better than another is difficult, especially when it relies on reuse. We can illustrate this with another example. At what point does it make more sense to reuse a ceramic or glass mug than constantly to use paper cups, once each?

Some would answer the ceramic mug, because it can be reused, whereas the paper cup is thrown away after (usually) one use. But ceramic is notoriously energy intensive to create (particularly the firing process). Mugs and glasses require a lot of energy in their creation—much more than a single paper cup.* In order to make a fair comparison, we need to measure all sorts of factors, such as how much hot water and soap is used to wash the mugs in between use. Fortunately, someone's already done this for us, as shown in the following table.** The important factor here is number of uses. It would take 70 uses of the ceramic mug to offset the water, energy, and materials used in the production of paper cups. So, for a single use, the paper cup is far better. In fact, for up to 69 uses, the paper cups (all 69 of them) would be better for the environment. But, at the 71st use of a ceramic mug (and everything after that), it would be better for the environment. This requires us, then, to assess where and how drink containers are used. Perhaps, for temporary, transient uses (like on an airplane or for take-out), paper is better after all. However, wherever possible, if mugs can be reused (as in offices, homes, restaurants, and so on), they are better still.

Paper Cup	Ceramic Mug	Glass
1 use	71 uses	37 uses

* Glass is less energy intensive and much more recyclable than ceramic.
** "Paper Versus Polystyrene: A Complex Choice." *Science* 251:504-505. Hocking, M. B. 1991

The lesson here is that, in order to determine which solution has less impact, we have to take into account how often it is used. Any dematerialization of an existing product may introduce issues that can only be accounted for by addressing repetitive use.

Other, similar, examples include disposable diapers versus reusable cloth diapers, and so on.

My mother (and probably yours as well) cares about the environment and wants to make good decisions when she purchases and uses things. However, like most people, she has neither the time nor the interest in becoming an expert in all of these disciplines just to make decisions in the grocery store. Nor does she have access to the data necessary to make better choices. This creates an important problem: if experts can't agree or determine, absolutely, which choices are better, how can we expect the rest of us to?

The only answer is that designers and developers, who are in a position to make evaluations based on privileged data and decisions based on deeper understandings, need to help by employing their skills at every possible step of the process. This is, perhaps, the most important contribution we can make in the world, but it's often unsung.

A Better Way?

The reality of most sustainability measuring is that there are no perfect scores. At best, there is "better" and "worse"—and these aren't often clear. Complex systems, by definition, connect to many issues and often create surprising interactions and conclusions and even unintended consequences. An example as seemingly simple as "paper" or "plastic" grocery bags yields unsatisfying results (both sides can be adequately argued), depending on how and what you measure.

So, as a designer, if you're looking for a cookbook to tell you how or what to design, I'm sorry that there is none. For most of your time on a project, you may only have a hazy idea that you're heading in the "right" direction of a more sustainable solution. You should be prepared to be blindsided by assumptions you thought were undeniable (such as plastic being better than glass or Styrofoam being better than paper in certain circumstances). Maintaining an open mind and a systems-level perspective can be difficult in the face of myriad issues and often trivial pronouncements in the press or even from experts.

You won't ever create a perfect solution. Ever. You will have to be satisfied with creating better solutions along the way—each update, hopefully, better than the rest, and potentially no solution ever reaching your ideal vision of "how it should be done." Every design solution is a compromise of some kind, bowing to structural, financial, or environmental realities, and conforming to customer, market, or client desires. That's the nature of design. If you're creating real solutions for real people, the market will probably not yet be ready for the ultimate solution you envision. This, of course, doesn't mean it's not worth trying to push as far as possible and to achieve as much as you can for the moment.

As long as you check your assumptions from time to time, keep educating yourself and others on the current state-of-the-art of both understandings and possibilities, and strive to do what you can, you're on the right track.

Get moving!

"Sustainability is the competitive strategy in boom time, turnaround strategy in down time, and survival strategy in a collapse."
                                        —Hunter Lovins

CHAPTER 3

What Are the Approaches to Sustainability?

There are several approaches, or frameworks, for understanding sustainability and its impact on the design and development of products and services, as well as on corporate behavior. For the most part, none is complete and only together do they form a full picture of criteria, processes, and approaches to more sustainable solutions. The proponents of some frameworks argue that theirs is complete or the best, but this is hardly true. However, all frameworks are lenses that contribute to the larger understanding of the issues, measures, and approaches of sustainability.

In this chapter, I've tried to discuss the most popular and influential approaches, but, no doubt, more will arise in the future. Each framework discussed is accompanied by a diagram summarizing its approach in the context of the expanded spectrum discussed in the first chapter. Hopefully, you'll find inspiration and something of value in all of these approaches. Don't let anyone tell you that there is only one way worth pursuing or from which to understand sustainability. Ultimately, you'll have to create your own model as a synthesis of everything meaningful you find in any of these frameworks, which is what I've tried to do for you in the "summary framework" at the end of this chapter

Whichever framework (or combination) you use, it's important to identify those that integrate well with your current development process and culture.

Natural Capitalism

Also known as eco-efficiency[1], the Natural Capitalism framework was developed by three luminaries of sustainability: Paul Hawking, Amory Lovins, and Hunter Lovins (see Figure 3.1). It is described in detail in their influential book 2002 *Natural Capitalism.* It is a framework for rethinking the value of social and natural resources in the context of business. Easy to understand, it creates a quick foundation for understanding the value of sustainability and the new perspectives around sustainable design and development.

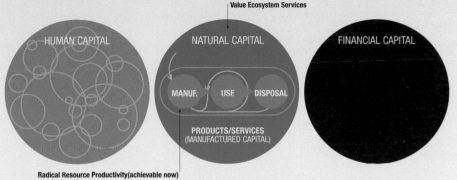

FIGURE 3.1
The Natural Capitalism framework.

Strengths: Clear, simple model of capital and value. Covers social, environmental, and financial issues. Business-relevant and friendly. Integrates well with design and business functions. Easy to address in the development process. Easily combined with other frameworks.

Weaknesses: Doesn't describe a development process. Not deeply detailed.

---

1   The term eco-efficiency was coined by the World Business Council for Sustainable Development (WBCSD) in its 1992 publication "Changing Course." Wikipedia.

At its heart, this framework describes four types of capital:

- **Natural Capital** describes the natural resources we get from the earth. These include physical things, such as materials and energy, as well as processes and effectives, such as ecosystem services and resiliency due to biodiversity.

- **Human Capital** is the value we get from the work and ideas contributed by people (individuals and society as a whole).

- **Manufactured Capital** describes the materials and energies people create through industry that aren't found in nature. This includes all intellectual property (IP), and it is where the bulk of value contributed by designers is found.

- **Financial Capital** is money and the many forms it comes in (such as cash, credit, stocks, bonds, options, and so on). Financial capital is an incredible invention that allows us to use and invest in the other forms of capital in more effective and innovative ways.

In this framework, four primary shifts are promoted:

- **Shift 1: Radical Resource Productivity** is the ability to increase dramatically the productivity of our use of natural resources—and, correspondingly, reduce the material and energy intensity of the products and services we create and use. Though it isn't specifically mentioned in the framework, this idea is easily extended to all forms of capital, not merely natural capital. Doing more with less has been a tenet of design, at times. It is responsible for the dramatic reductions in energy use per capita in the state of California over the past 30 years. For example, refrigerators use about a fourth the energy they did three decades ago, and homes use much less energy for heating. This shift is merely the continuation—and acceleration—of what we're already so good at doing (designing efficient solutions). Included in this shift is a reduction of toxic materials throughout the product life cycle.

- **Shift 2: Ecological Redesign** shifts our perspectives and processes to biologically-inspired models. Biomimicry describes one such approach. By better understanding nature, which has been creating and improving solutions for millions of years, we can create better solutions that efficiently use all four forms of capital.

- **Shift 3: Service and Flow Economies** shift the emphasis from products to services and from objects to outcomes. Instead of concentrating on solutions-as-objects, we can often deliver better function, value, and relationships from services—even if there is an object involved in some way. Solutions like rental cars and downloadable music deliver better efficiency, as well as new functionality, than the product-oriented models they replace (in these cases, everyone owning his own car or music CDs).

- **Shift 4: Investing in Natural Capital** builds a stronger resource base, as well as a more resilient world. The authors of *Natural Capitalism* often remind us that we use natural resources as if they are infinite and that our development in the world acts as if the world itself doesn't matter.[2]

Natural Capitalism clearly convinces us that we can have a huge effect on the efficient use of all forms of capital now, using technologies we already know well. We don't have to wait for new advances to make a radical difference, although there are many wonderful advances on the horizon that will only improve things further. Instead, we simply need to make efficiency a priority. Natural Capitalism prioritizes the most important sectors of energy and materials use, like transportation, home heating, electricity generation, water use, and food, in order to make the most gains as quickly as possible. However, it also notes that all pollution is inefficiency at work, so while it triages around the most critical issues, it's also inclusive of most of the issues important in the other frameworks.

We don't have to wait for new advances to make a radical difference ... we simply need to make efficiency a priority.

---

2   Lovins, Hunter, *Development as if the World Matters*, World Affairs Journal, 27 June 2005, www.natcapsolutions.org/publications_files/WAJ_May2005.pdf

Critics of Natural Capitalism point to its failure to challenge some basic tenants of production, consumption, and waste that should be rethought. For example, proponents of the Cradle to Cradle framework (described next) would still characterize Natural Capitalism as a "cradle to grave" approach, meaning that while efficiency may go up and waste will be lowered, there is still waste of some kind (and any amount is too much). The idea of waste doesn't appear, for example, in nature. The outputs of one process are the inputs of another. This is one of the fundamentals of Biomimicry (see the framework that follows). Furthermore, it is possible to develop manufactured capital that can be fully recycled into materials of like quality when they reach the end of their usefulness. This differentiates waste or "cradle to grave" thinking from "cradle to cradle" thinking. (See the next framework.)

Nonetheless, it's not likely that our society will suddenly jump from deeply-ingrained "cradle to grave" solutions to revolutionary "cradle to cradle" solutions in one step. The progress will be gradual, and Natural Capitalism represents an important framework for improving efficiency and effectiveness on the road to deeper frameworks, like Cradle to Cradle.

Many Types of Capital

Natural Capitalism's four types of capital are easy to understand, but they aren't the only model describing forms of capital other than financial. There are several lists of capital you may encounter and while some list four, five, seven, or even nine types, all are compatible in that the same basic categories are simply broken into finer subcategories in some:

Many Types of Capital (continued)

Natural Capitalism	Progressive Economics	Michael Fairbanks	LASER Manual
Natural	Natural	Natural Endowments	Natural
Human	Human Social	Human Cultural Institutional	Human Social Cultural Institutional
Manufactured	Manufactured	Humanly Made Knowledge Resources	Built Technological
Financial	Financial	Financial Resources	Financial Technical Exchange

As you can see, all of the forms of capital from every system include National Capitalism, Financial Capitalism, and several different categories of Human Capitalism.

The difference in numbers of types of capital can be accounted for in the system's differentiation of capital. To understand some of the finer subcategories of capital, use these descriptions:

- Natural Endowments (location, soil, assets, forests, climate, water, etc.)

- Financial Resources (savings, reserves, etc.)

- Humanly Made Capital (buildings, infrastructure, roads, telecommunications, assets, etc.)

- Institutional Capital (legal protections, intangible property, shareholder value)

- Knowledge Resources (patents, IP, etc.)

- Human Capital (skills, insights, capabilities)

- Cultural Capital (music, language, ritual, attitudes, perspectives, values, etc.)

Many Types of Capital (continued)

Forms of Capital:

Natural Capitalism
- Natural
- Human
- Manufactured
- Financial

Progressive Economics
- Natural
- Human
- Social
- Manufactured
- Financial

Michael Fairbanks: Changing the Mind of a Nation Elements in a Process for Creating Prosperity
- Natural Endowments (location, soil, assets, forests, climate, water, etc.)
- Financial resources (savings, reserves, etc.)
- Humanly Made Capital (buildings, infrastructure, roads, telecommunications assets, etc.)
- Institutional Capital (legal protections, intangible property, shareholder value)
- Knowledge resources (patents, IP, etc.)
- Human Capital (skills, insights, capabilities)
- Cultural Capital (music, language, ritual, attitudes, perspectives, values, etc.)

LASER Manual
- Social Capital
- Cultural and Historical Capital
- Human Capital
- Institutional Capital
- Financial Capital
- Potential Exchange Capital
- Built Capital
- Technological Capital
- Natural Capital

Cradle to Cradle

Also known as eco-effectiveness, Cradle to Cradle, or C2C, is a popular framework that demands significant change, for good reason, but represents a high bar to reach for most organizations (see Figure 3.2). It is a powerful perspective on the cyclic nature of waste and food, as well as the need to keep technical and biological materials separated. Great strides have been made under the C2C framework, but it takes commitment and support from the highest levels of an organization to achieve. A new, accompanying certification includes more detailed criteria.

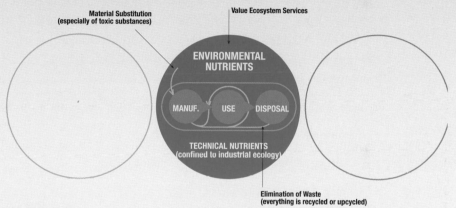

FIGURE 3.2
The Cradle to Cradle framework.

Strengths: Business-relevant and friendly. Integrates well with design and business functions. Easy to address in the development process. Easily combined with other frameworks.

Weaknesses: Doesn't cover financial or social issues. Doesn't propose metrics. Doesn't describe a development process. Not deeply detailed. Doesn't favor natural or organic materials and can be biased toward technological solutions. Doesn't explicitly value local production, transportation, product life span, or embodied energy of components.

Another popular framework for understanding sustainability was proposed by Walter Stahel in the 1970s and made popular by William McDonough and Michael Braungart in their 2002 book, *Cradle to Cradle*. These two (one an architect, the other a materials engineer) have built solutions that have surprised and amazed their clients, industries, and the public. For example, they specialize in redesigning the manufacturing process and focusing on materials and process selection in order to sharply reduce toxicity (or eliminate it altogether) throughout the life cycle of products, but particularly in manufacturing. These solutions often result in factories that emit cleaner water after manufacturing than what was used to begin with.

Think about that for a moment. Imagine a factory whose outputs are cleaner than its inputs. That's more than merely a factory—it's now a system for cleaning and restoring the water or air or other aspects of the environment. Think about how that one alteration could change everything we expect about a factory. That's what McDonough and Braungart excel at.

The basis for the Cradle to Cradle (C2C) approach involves four principles:

- Elimination of hazardous (toxic) materials

- "Waste equals food" (changing our definition of "waste")

- Use the current solar income of energy

- Use "upcyclable" materials

The Cradle to Cradle framework (also known as *eco-effectiveness* to differentiate it from frameworks that focus on efficiency) requires us to question all materials and eliminate (or at least drastically reduce) all toxic materials. This framework has little tolerance for materials that harm customers, employees, workers, or the environment. Braungart and McDonough have found that a thorough examination of manufacturing materials and processes reveals no need for many of the toxic chemicals that are assumed necessary, out of habit or legacy. For example, industry malaise, assumptions, and lack of innovation in the carpet business fed the use of several toxic materials and processes until Interface hired Braungart and McDonough to examine their business. These consultants found that they could recommend the elimination of many assumed "necessary" chemicals with little or no impact on the manufacturing process and the elimination of many others with changes in some processes. The result is far fewer toxic materials used in the manufacturing process, and a product that can be recycled with even fewer toxic chemicals used in the recycling process.

The Cradle to Cradle framework, like many others, acknowledges the need to address the entire life cycle of production, transportation, use, and disposal, as well as the need to foster diversity in the environment. C2C differentiates between *biological nutrients* and *technical nutrients*. Technical nutrients are materials not found in nature that should not be put back into nature, whether benign or not. Instead, they should be recycled into other technical nutrients. Likewise, biological nutrients can be taken from nature and put back into nature, if in acceptable forms. For example, although paper is made from natural materials and can biodegrade under certain circumstances, paper laden with toxic dyes shouldn't simply be buried in the ground.

This distinction is important and highlights the need not to mix the two in one material (called a *monstrous hybrid*), since the combination can't be recycled easily into either system. This combination of products represents one of the biggest problems with modern packaging. Take, for example, the ubiquitous juice box or milk cartons with plastic screw-top spouts. Because these packages are made from dissimilar materials bonded together that are difficult to separate (just try peeling the aluminum foil from the paper carton and the plastic laminate), they're essentially impossible to recycle and must be thrown way.

Cradle to Cradle casts the manufacturing process and product use in terms of metabolic material flow, meaning manufacturing can be compared to the natural metabolic processes of plants, animals, fungi, and bacteria. When approached this way, we quickly see a major distinction between human production and natural production: nature has no waste.

Any given process or organism in nature may expel materials, but it's never considered *trash*. Everything expelled is food for some other organism and process. Not so for manufactured human products. When we're done with something, we often just bury it—sometimes in a way that retards or even makes any kind of breaking down or reuse impossible. We treat the environment not only as an unlimited store where everything is never-ending and practically free, but also as an unlimited trash dump.

This simple new understanding can have remarkable outcomes. If everything can be made into food for some other process, then recycling can be completely rethought. In addition, recasting waste as food increases efficiency and often lowers costs, much to the delight of companies. It doesn't correct problems of over-consumption and over-consumerism, but it does start to deal more realistically with their effects.

Once we re-imagine a production process without waste, we can extend that concept to the use of the manufactured products as well. The outputs of use can also be reconceived as inputs for other processes, thus reducing further the need to dump low-quality material into landfills or the ocean. At this point, the entire life cycle of a product, from sourcing raw materials and refining them, to manufacturing and distribution, to use and finally to disposal, becomes a potential closed-loop cycle producing usable outflows at all points.

> **We treat the environment not only as an unlimited store where everything is never-ending and practically free, but also as an unlimited trash dump.**

In addition, careful engineering and ingenuity in materials manufacturing and use can expose opportunities for materials the authors describe as "upcyclable" rather than merely recyclable. While a recyclable material can be used in some other form, it's often only reusable in a degraded, less valuable form because it has lost important aspects of its quality (such as purity, longevity, or structural integrity). White paper, for example, is usually only recycled into lower forms of paper, such as newsprint or cardboard, but not back into premium, white, high-grade paper. This is due to the paper recycling and manufacturing processes that stress paper fibers, causing them to shorten and weaken with each processing.

While some recycled materials must be constantly downgraded in quality (such as paper), others can be reused at the same level, remanufacturing one product or material into another of similar quality. This process happens currently with certain types of glass and metal. Aluminum parts, if appropriately separated, can be melted down and recast into new parts with the same characteristics of virgin aluminum. The same is often done with other metals. For the most part, this is possible with glass, although not all glass is alike. If glasses of different types are melted and mixed (brown and green bottles with clear window glass, for example), then they cannot be easily separated and remade into pristine, clear glass. Such a mixture may only be downcyclable into brown glass bottles.

> **If everything can be made into food for some other process, then recycling can be completely rethought.**

An upcyclable material, on the other hand, is one that can be not only reused again with the same material characteristics, but can also be reclaimed and improved to have better characteristics than the original material. In effect, it can be pushed up the value chain and improved. Not many materials can yet be upcycled (aside from many types of glass and aluminum and a couple types of plastics), but more are being developed.

Cradle to Cradle focuses on effects and efficacy instead of whether a substance or process is "natural" or not. It doesn't give preference to natural or organic materials, since nature produces many materials that aren't healthy for humans or the ecosystem (such as uranium and arsenic). This tends to bias the framework toward technological solutions, in particular, newer advanced technologies, instead of traditional approaches that often avoid high-tech solutions (like genetic engineering). Cradle to Cradle also doesn't deal with the cultural or systemic implications of consumption, nor does it discuss how we might limit it.

Cradle to Cradle is also now a certification system, although it can be expensive to apply and certify. The high-level criteria used for certification include the following:

1. Product/material transparency and human/environmental health characteristics of materials

    (including toxic materials and processes)

2. Product/material reutilization

    (including use of recycled materials and well-defined material recovery, such as a take-back program)

3. Production energy

    (including the use of renewable energy for product manufacturing and product uses)

4. Water use at manufacturing facility

    (including the implementation of conservation and discharge measures at the manufacturing plant)

5. Social fairness/corporate ethics

    (including whether the organization provides a CSR and makes other decisions transparent and whether these materials are audited by a reputable third party)

More information on the certification criteria is available on the Web site: **www.mbdc.com/c2c/**.

Ford Model U

In 2003, Ford created a concept car, the Model U, in order to explore issues of sustainability throughout the automobile manufacturing, use, and disposal processes (see Figure 3.3). Working together with William McDonough and Michael Braungart, Ford's designers and engineers were able to take a fresh look at materials, processes, and solutions for automobiles. While this is only a concept car, and there are questions about whether or not the advances explored are making their way toward production in any real Ford cars, the Model U is useful as an example of cradle to cradle thinking applied to complex industrial products.

FIGURE 3.3
Ford's Model U concept car.

Powered by a hydrogen fuel cell (but, remember, this isn't a real production car), the Model U project projects 25 percent more fuel efficiency over similarly-sized cars and nearly zero emissions. But that's just the start. The real advancements come in the materials and processes developed for this concept, which include the following:

- Soy-based foam in the seats

- Soy-based and aluminum body panels

- Sunflower seed-based lubricant

- Fabrics for seats and dashboard are made from polyester that is 100 percent recyclable

Ford Model U (continued)

- A foldable canvas roof made from PLA (polyactide), a bioplastic fabric derived from corn that can be safely returned to the biological nutrient flow

- Fuel-efficient tires made partially from corn-based fillers (making them lighter with less rolling resistance)

- Interchangeable components with fewer variations for easy assembly and disassembly, as well as reconfiguring components later to enhance the car's capabilities

- Modular manufacturing so that several variations of the car can be made on the same factory assembly line

- Paint cured with ultraviolet light instead of high temperatures

Concepts like the Ford Model U not only serve the purpose to prototype new solutions, but they also send a signal to customers, suppliers, partners, and other stakeholders about the possibilities inherent in cradle to cradle thinking.

Biomimicry

Like the Cradle to Cradle perspective, *Biomimicry* takes its inspiration from natural processes (see Figure 3.4). Promoted by Janine Benyus, Biomimicry isn't so much a framework as it is an approach to re-imagining the design and development process. It is a perspective that searches for new ways of creating sustainable materials, products, services, and other solutions by learning how nature already works.

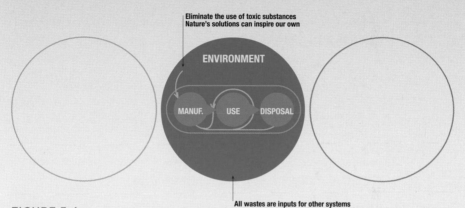

FIGURE 3.4
The Biomimicry framework.

Strengths: Inspirational. Focused on nature's accomplishments as both model and source of solutions.

Weaknesses: Incomplete, subjective metrics. The Design Spiral is a powerful procedural approach but is better at integrating into traditional development processes rather than replacing them.

Natural processes are incredible. The materials and processes already at work in nature rival and in many cases far surpass the complexity humans have managed to create. Janine Benyus is the founder of this perspective and the author of an excellent book by the same name, *Biomimicry*. In it, there are many examples of natural materials created by plants, animals, and other organisms that hold great promise for human solutions. As described previously, nature has no waste. In addition, some materials in nature rival what we've been able to create with our technical solutions. For example, silk is stronger than steel. Nature already has glues stronger than the best we've created and a host of other solutions we have much to learn from. Again, nature's solutions are more efficient, more sustainable, and create no waste. Nature can serve both as a source of inspiration and a source of materials and processes that we can use and emulate to create better, more sustainable solutions.

Biomimicry asks questions of designers and developers that open us to new answers:

- How does life make things? (Humans currently rely on heat, beat, and treat to solve problems.)

- How does life make the most of things? (Nature often adds information/data to its solutions to make them perform better.)

- How does life make things disappear things into systems? (Nature's solutions aren't isolated from their contexts.)

In addition, Janine Benyus has observed several principles of nature that can be effective guides to help us find new solutions:

- **Self-assembly** (Many of nature's solutions, like crystals and DNA, self-assemble without the need for factories.)

- **Solar transformation** (Many of nature's solutions rely on the sun to power themselves.)

- **Power of shape** (The shape of molecules, organs, and organisms often powers nature's solutions.)

- **Color without pigments** (Several of nature's solutions use thin film interference to create color; for example, a chameleon's ability to change its color to that of its surroundings, rather than its pigment.)

- **Cleaning without detergents** (Leaves, for example, self-clean because of their surface texture, repelling dirt and water.)

- **Water-based chemistry**

- **Metals without mining** (Plants and fungi extract metal from their environment without resorting to strip mining or methods that scar the earth.)

- **Green chemistry** (Many of nature's solutions are able to surpass ours without toxic chemistry.)

- **Timed degradation** (All of nature's solutions disintegrate at some point, leaving building materials for other processes instead of waste.)

- **Sensing and responding**

- **Growing fertility**

- **Life creates conditions conducive to life** (Instead of toxic conditions that destroy life.)

- **Decentralization and distributed control** (Resilient solutions are often decentralized.)

- **Simple building blocks** (Create deep complexity.)

- **Use of feedback loops** (Influence, rather than control.)

- **Redundancy**

- **Cyclic solutions**

- **Diverse solutions**

The Design Spiral

In the words of the Biomimicry Guild, nature can be "model, measure, and mentor."[3] This means that nature can be used as a guide to translate functions into biological terms. We can strive to discover and emulate nature's processes and materials, evaluate solutions against nature's own principles, and learn from nature as a source of information and strategies, not merely materials.

---

3 www.biomimicry.net

Because Biomimicry doesn't suggest a tool set as much as it represents a way to think about and judge new solutions, developers might find it difficult to use Biomimicry as anything other than a source of inspiration for new solutions or new materials. Often, the framework is more helpful in formulating the processes of research and development of new materials and manufacturing processes rather than finding solutions to human needs. However, Biomimicry's framework for the development process, the Design Spiral (see Figure 3.5), developed by Carl Hastrich in cooperation with the Biomimicry Guild, is a development process that maps roughly to contemporary design processes but from a biomimetic perspective.

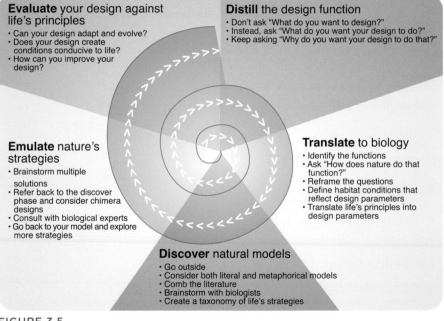

FIGURE 3.5
The Design Spiral.

The Design Spiral from the Biomimicry Guild isn't terribly different than the process many designers already enage in when designing new solutions. However, the order is somewhat different. Where contemporary design approaches suggest research, then prototyping, then evaluation, the Design Spiral expands the question of design brief through translation into similar biological processes (see Figure 3.6).

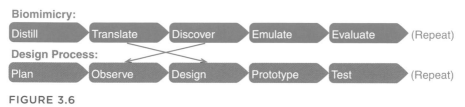

FIGURE 3.6
Expanding the design process into similar biological processes.

Biomimicry teaches us to evaluate design solutions by comparing them with nature's principles and processes. In this way, it's an addition to the design process but not a replacement for it, since nature's principles don't reflect human requirements (either on an individual or societal level). So Biomimicry is a framework that should be used in combination with design processes, not instead of them. In fact, Biomimicry fits best into part of the design and prototype phase since the objectives of design research and testing are focused more appropriately on evaluating human values and not natural values.

A more appropriate process that combines the two might be seen in Figure 3.7.

FIGURE 3.7
Combining the two processes into one.

> **Biomimicry teaches us to evaluate design solutions by comparing them with nature's principles and processes.**

Developers using any contemporary development process should find it comforting to know that their current processes—especially other spiral-based processes such as Extreme Programming—can work within the Biomimicry framework with little modification. Instead, a shift in perspective and priorities is most important. Throughout any process, questions of the environment and nature's solutions can help focus developers on critical issues, inspirations, and possible solutions already existing in nature. For example, it's already common that designers carefully cast new product specifications in terms of human needs and outcomes. However, the Design Spiral also asks us to question the brief from nature's perspective: "How does nature accomplish this?" and "What are nature's needs?" The Design Spiral asks us to recast the location of solutions in nature's terms as well—habitat, climate, nutrients, and so on.

Biomimicry doesn't offer much guidance in terms of social or financial sustainability. It is primarily focused on environmental impacts. Indeed, nature isn't a system that reflects human values so it can't be used to guide decisions made about them. However, even in nature, there may be answers to social and financial models of sustainability. In fact, throughout the entire development process, biomimicry reminds us to "look for answers already resolved in nature."

...throughout the entire development process, biomimicry reminds us to "look for answers already resolved in nature."

PAX Scientific

An impressive example of Biomimicry resulting in superior solutions involves the products from PAX Scientific and its subsidiaries. These seemingly simple impellers take their form from carefully observing vortexes in nature (see Figure 3.8). PAX's founder, Jay Harmon, and his scientists designed the complex, multiple-axis shape of their impellers around the most efficient geometries they found in biological systems (including plants and animals). As a result, their pumps dramatically increase the amount of water or air they move or mix (depending on the application), increasing efficiency and reducing power consumption, noise, drag, and costs. In addition, the designs are carefully optimized for manufacturing to reduce material use and waste. Their products have been so efficient and successful, that PAX is applying these designs to every imaginable use of moving fluids, from automobiles to laptop computers to wastewater management. **www.paxscientific.com**

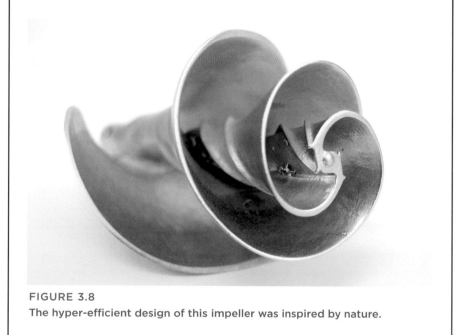

FIGURE 3.8
The hyper-efficient design of this impeller was inspired by nature.

Life Cycle Analysis

The most exacting and accurate framework for assessing solutions is Life Cycle Analysis (LCA), an entirely quantitative approach (see Figure 3.9). There are several variations of LCA tools, but conceptually they are largely the same. LCAs are usually expensive, time-consuming, and difficult (if not impossible) to perform, but they deliver the most accurate and useful evaluation of materials and energy use.

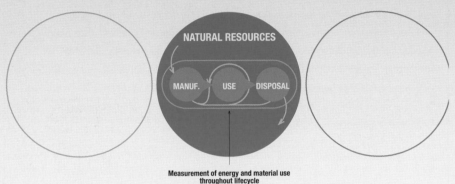

FIGURE 3.9
The full life cycle of a product or service includes many phases.

Strengths: Comprehensive, objective, easier to measure for existing products and services, rather than for proposed ones.

Weaknesses: Doesn't adequately address financial or social analyses. Difficult, time-consuming, and costly to perform. Most of the data needed for adequate evaluations aren't available from organizations. Can't be adequately performed in the design and prototype stages of development.

According to the U.S. Environmental Protection Agency, LCA is "a technique to assess the environmental aspects and potential impacts associated with a product, process, or service by using the following criteria:

- Compiling an inventory of relevant energy and material inputs and environmental releases

- Evaluating the potential environmental impacts associated with identified inputs and releases

- Interpreting the results to help you make a more informed decision"4

This requires the examination of the materials and energy consumed and emissions produced at each stage of a product's life, from the procurement of raw resources through manufacturing, transportation, and selling, throughout use, and finally to disposal and recycling (see Figure 3.10). (Ideally, everything would be recycled completely.)

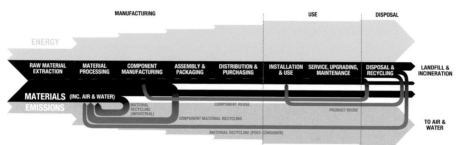

FIGURE 3.10
The full life cycle of a product or service includes many phases.

Every choice in a product's development affects its impact. These can include process choices (such as where and how to manufacture a component), transportation choices, and material choices (what each component is made from). For example, transporting products great distances via train is usually more efficient (and has less impact) than transporting them via truck or airplane. Also, some materials, like aluminum and stainless steel, have far bigger impacts than PET plastic and paper. Wise choices during development can make a big difference in the resulting environmental impact.

---

4 Curran, Mary Ann, "Life Cycle Assessment: Principles and Practice," National Risk Management Research Laboratory, Office of Research and Development, U.S. Environmental Protection Agency, Cincinnati, OH, May 2006, www.epa.gov/nrmrl/lcaccess/pdfs/600r06060.pdf, posted 3/20/2008

Energy, materials, and other inputs are consumed at each step of a product or service's life cycle. Also, at each step, emissions to the air and water, plus material waste, are generated along with the intended products and components. These all weigh into the impact. Even seemingly simple products can create deeply complicated LCA requirements.

A thorough LCA looks at the process and location for sourcing raw materials (including mining, processing, purification, and transportation). Metals, in particular, often require processes that heat raw materials several times, casting them into some other form (such as slugs or ingots), transporting them to other facilities (requiring storage, distribution, processing, or manufacturing), reheating them and recasting them into parts, and then transporting them again in a lengthy cycle. The simplicity we think we see in finished parts (or intend in our designs) often masks an even greater complexity in process, as well as material and energy use, that may total from 30–50 percent of the product's impact.

Packaging, too, represents a surprisingly high percentage of the materials and energy impact for many products. Consider the packaging costs for a bottle of perfume—the energy and materials that go into manufacturing the bottle, the label, and the series of boxes it will be shipped in can be staggering compared to the perfume contained within.

Even seemingly simple products can create deeply complicated LCA requirements.

When a customer purchases an item, it moves from the manufacturing category to the use category. For some products, the impact during the use phase may be small (such as for furniture). For others, however, the majority of their impact may occur during this phase (such as appliances and clothes).

But this still isn't the whole story. When products are disposed of, they require more material and energy to collect, recycle, and dispose of. Some, in fact, have as big an impact in this phase as in any of the others. Nuclear waste and CRTs are notoriously difficult to dispose of appropriately, generating greater-than-average materials and energy impacts in doing so (mostly due to their toxicity). Recycling materials is a good idea, but it also requires energy and materials for collection, sorting, reclaiming, cleaning, and transportation. These, too, must be factored into the equation.

Using recycled material or energy can often create a positive impact on LCA calculations, lowering overall material and energy impacts. However, this only occurs for the recycled material or energy that is actually used in the manufacturing phase, not for the potential recycling that might occur because a product is designed to be recycled. In other words, just because that PET bottle *can* be recycled into other bottles or other products, doesn't mean it *will* be. It could, just as easily, be dropped into the waste stream, end up in a landfill, or float endlessly in the ocean for years. The only recycled content that counts is that which gets recycled.

As daunting as an LCA seems to be, it is possible to conduct one, and there are two basic approaches.

> ... just because that PET bottle *can* be recycled into other bottles or other products, doesn't mean it *will* be.

Process-based LCA, developed by the Society for Environmental Toxicology and Chemistry, the U.S. EPA, and the International Organization for Standardization (ISO) breaks down the process for producing a product into its constituent activities, and evaluates the environmental impact of each individual step. The diagram in Figure 3.11 shows the hierarchy of process activities that must be identified in a process-based LCA. For each process in the diagram, materials and energy used, as well as emissions produced, must be identified, and the totals rolled up to create a full picture of the environmental impact of a product. These models require a considerable amount of work to develop, since they can require reaching far back into a product's supply chain to collect relevant data. Because systems are interconnected, it's not always clear where to consider a process's boundaries, when to stop collecting data, and how to collect the data in similar formats so that all of the data can be used in the same assessment.

OVERALL PROCESS

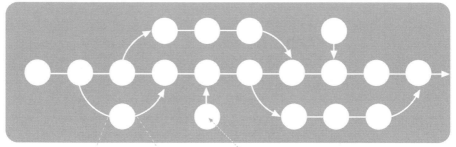

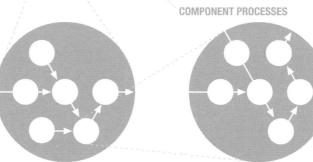

SUB-SYSTEM   **SUB-SUB-SYSTEM**

FIGURE 3.11
Hierarchy of process activities in process-based LCA.

Economic Input-Output LCA (EIO-LCA) is an attempt to simplify this arduous process by focusing on the likely inputs and outputs that will have the most impact. Rather than measure actual materials and energy from the product or service's life cycle (which is often unknown during development), designers and engineers use proxy data from common sources that average the impacts (see Figure 3.12). Developed by Wassily Leontief, who won the Nobel Prize in 1973, it represents a "general interdependency" model. This drastically simplifies the process, both in time and expense, but delivers estimates that are not as accurate. Nonetheless, it's the only way to approach an LCA during the design and prototyping processes, when a finished, actual solution doesn't yet exist. Data is available publicly at Web sites like **www.eiolca.net**.

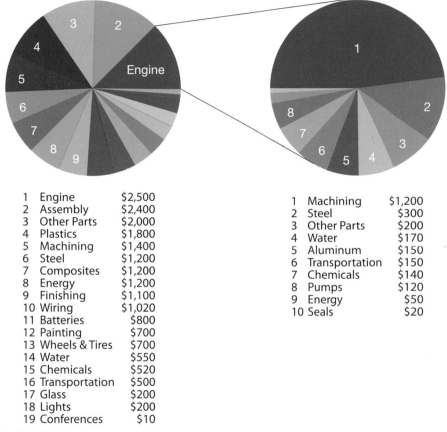

1	Engine	$2,500
2	Assembly	$2,400
3	Other Parts	$2,000
4	Plastics	$1,800
5	Machining	$1,400
6	Steel	$1,200
7	Composites	$1,200
8	Energy	$1,200
9	Finishing	$1,100
10	Wiring	$1,020
11	Batteries	$800
12	Painting	$700
13	Wheels & Tires	$700
14	Water	$550
15	Chemicals	$520
16	Transportation	$500
17	Glass	$200
18	Lights	$200
19	Conferences	$10

1	Machining	$1,200
2	Steel	$300
3	Other Parts	$200
4	Water	$170
5	Aluminum	$150
6	Transportation	$150
7	Chemicals	$140
8	Pumps	$120
9	Energy	$50
10	Seals	$20

FIGURE 3.12
Averaging the financial impact of a product.

LCA is very much an engineering framework and can be cost- and time-prohibitive for smaller organizations and individuals to complete. In addition, the more complex the solution, the more difficult the assessment is, often in an exponential relationship to the number of parts and materials.

In addition, the realities of manufacturing and distribution make the reporting difficult, if not impossible. For example, the same product produced in San Diego will have a drastically lower transportation impact when used locally than when distributed to San Francisco, Denver, and Portland, Maine. In order to accurately portray this impact in labeling and in order to make informed decisions (for both manufacturers and customers), each location of consumption would have to be calculated and displayed separately. For contemporary products created in multiple locations, shipped all over the world, with switching supply chains at the whim of last-minute market changes, this becomes nearly impossible.

Not only are LCAs time-consuming, but most manufacturers and service providers also don't yet record or report the minute data needed to make these deep assessments possible. Current accounting methods for manufacturing, distribution, and maintenance operations just don't require this data, and as such, most organizations don't see the need to pay attention to it. This may change in the future, especially as organizations feel the need to increase their efficiency and effectiveness. But, for now, LCA input data simply isn't available for the vast majority of solutions.

Luckily, there are, at least, a number of software tools (such as, **www. gabi-software.com, www.ecoinvent.ch,** and **www.simapro.com**) available to enable LCAs, and more are being developed all the time. Some tools are more geared toward specific industries, such as the Pharos system for buildings developed by the Cascadia Region Green Building Council (**www.pharosproject.net**). Likewise, the Leadership in Energy and Environmental Design (LEED) Green Building Rating System (U.S. Green Building Council 2008) also provides guidance to new building designers to improve the environmental performance of buildings (**www. usgbc.org/leed**). However, most industries are sorely lacking in tools specific to their needs.

While LCA tools can help developers roughly assess the environmental impacts of their designs, there are still issues that aren't easily addressed by integrating these tools, such as quality, durability, and consistency. For example, reducing the wall thickness of a part might reduce its impact but might also drop its structural integrity beyond quality targets. Substituting a different material might make the part stronger but will create a completely different impact. These are the choices (and the attention to detail) required of designs and engineers if they hope to reduce the impact of the solutions they design.

How important is life-cycle analysis? Let's use an example to illustrate in the following sidebar.

Which Is Better for the Environment—the Toyota Prius or the Hummer H2?

This sounds like a ridiculous comparison, doesn't it? Most people (as well as common sense) would tell us that the Prius is definitely, even unquestionably, better than the Hummer H2 (see Figure 3.13). However, how do we actually know this? On what basis are we making this assessment? I've used this example because it's a famous controversy in the sustainability world. As we found with the bag example in Chapter 2, even a one-piece product can bring up surprising issues. Consider how much more complex cars are.

FIGURE 3.13
Which is better for the environment? It's not always clear.

To do an adequate evaluation, it's important to look at the life cycle of these products. For example, in the use phase of each (in this case, driving), it's obvious that the Prius is better because it gets significantly better mileage (45 mpg for the Prius, 17 mpg for the Hummer). It doesn't get much clearer than that.

But what happens if we look at the phases of each, namely, the manufacturing phase, (including sourcing raw materials and components, assembling them—often in different places—and transporting them to buyers) and the disposal phase (what happens to them when we're finished with them, including any recycling of their components). Miles per gallon won't help us here. How do we measure these impacts and where do we find the data?

This presents another critical challenge: customers don't have access to this data. Again, we need to rely on experts, but in this case, experts don't have this data either. The only place to get the data (and it's incredibly complex data) is from the manufacturers themselves, and for the most part, they don't track this information. Even when they have partial data, they certainly aren't sharing it. So where does that leave us?

Industry experts can estimate this data and probably get reasonably close at figuring out which is more cost-effective, but this is conjecture, not real data. And, when experts do estimate, we have to question what their intentions are. In the Prius vs. Hummer example, one industry group did just this, to controversial conclusions.[5]

This particular study showed that, when you take into account the entire life cycle of each car, the Hummer's impact is less than the Prius. How can that be?

In the manufacturing phase, the Prius (and all other hybrids) suffer from increased complexity. Hybrids have two drivetrains and many batteries that both add significant weight and are manufactured from more toxic materials. In addition, hybrid technologies are new and haven't benefited from as much efficiency derived from design and manufacturing experience. The Hummer (and many SUVs) may actually have an advantage in this phase.

In the transportation phase, neither vehicle probably has an advantage over the other for cars sold in the U.S. Both travel considerable distances (the Prius from Japan, at the moment) and the Hummer from Indiana. To really make a determination, we'd have to look at the methods of transportation *and* the distances traveled. Ships are efficient but incredibly dirty. Trains and trucks a little less dirty (though not by much) but also less efficient per car. We'd have to take into account the types of fuel used as well. For now, let's just call it a wash.

---

5   A "Dust to Dust" study by CNW Marketing Research of Bandon, OR
    http://cnwmr.com/nss-folder/automotiveenergy/

Which Is Better for the Environment—the Toyota Prius or the Hummer H2? (continued)

In the use phase, the Prius clearly wins. Even though it carries around two drivetrains, it's still lighter and its regenerative breaking captures otherwise lost energy. Score a big one for the hybrid.

However, once we get to the disposal phase, things flip around. Hybrid batteries use highly toxic chemicals that must be disposed of carefully. For those that get junked properly (or will when they start to get old), the costs are much higher than for conventional cars. Proper disposal or not, these chemicals represent a higher cost, both financially and environmentally.

To start, although the Prius excels in the use phase, there are questions about the manufacturing and disposal phases. The issues look something like Table 3.1.

TABLE 3.1

Comparison of Two Vehicles

Phase	Prius	Hummer
Manufacturing	• Manufactured in Japan (for now) and shipped to the U.S. • Newer technology (not yet optimized) • Two engines and drivetrains (one, gas, one electric) requiring many more parts • Many batteries (which are highly toxic and require a lot of platinum and other metals)	• Manufactured in the U.S. (less transportation required) • Older technology • One engine and drivetrain (though large) • One battery
Use	• 48/45 2008 EPA mpg (city/highway)	• 11/17 2008 EPA mpg (city/highway)
Disposal	• More toxic materials • Replacement after only 100,000 miles?	• More material overall • Replacement after ~300,000 miles

This last point, how many miles to rate each car at, is the focus of the controversy. It's undeniable that the Prius excels in the use phase. It is probably a tie in the manufacturing phase (a case could be made that either is better, based on material amounts or types and the distance they travel). However, the study in question assumed the Prius would only last 100,000 miles based on the ratings on the battery (implying that the entire battery pack, a critical component, must be replaced after 100,000 miles in order to continue achieving the performance in its use phase).

Many people point to this as being unfair,[6] and, in fact, if the same numbers published in the report were amortized over 300,000 miles, just like the Hummer, the final impact scores would be very similar. In itself, this is surprising to most people who don't know about the mechanics of hybrid systems. Hybrids like the Prius are a kind of compromise. Essentially, they are an electric car with an on-board power generator that runs on gas. It's a good compromise, based on our current expectations of car performance and the current state of technology, but it's still a complex compromise that's anything but ideal.

Setting aside the issue of complexity, we don't really know yet how to rate the longevity of the Prius. Surely, the body and engine should last as long as the Hummer. But how do we rate the batteries (which may last longer than 100,000 miles but surely not as long as 300,000)? Toyota's response to the study was reactive and defensive and, rightly, challenged many of its assumptions, but it didn't illuminate any of the questions, and it didn't provide any data with which experts could recalculate the scores. If you read the response to Toyota's response,[7] written from the organization who created the study (funded by the U.S. automobile industry, by the way), it makes some important points that need to be addressed but it doesn't make the question (or answer) any clearer.

The only way to know for sure is to have access to the data and let a lot of knowledgeable people fight over the assumptions and projections. The problem is that no one but the

---

[6] A rebuttal from the Pacific Institute: www.pacinst.org/topics/integrity_of_science/case_studies/hummer_versus_prius.html

[7] CNW response: http://cnwmr.com/nss-folder/automotiveenergy/

Which Is Better for the Environment—the Toyota Prius or the Hummer H2? (continued)

manufacturers has access to the data and, likely, they don't even track the data to the level required to do a full-scale LCA. This is a serious problem because it destroys the possibility of claims on either side being validated. In essence, the experts aren't able to tell us which is better.

There are other circumstances as well. For example, the Prius sales figures are sending a powerful message to the car industry about how interested drivers are in environmentally better solutions. That alone may be worth any difference in environmental impact.

It's probably safe to say that the Hummer is worse for the environment than the Prius. (Although this doesn't take into account other benefits buying a Prius may have, like the signal it sends to the automobile industry about environmental concerns on the part of customers.) However, the difference is probably not great, and it's probably also safe to say that a non-hybrid five-person sedan, like Toyota's own Corolla or Camry, may actually be better for the environment over its entire life cycle than the Prius. Even smaller cars, though not hybrids, like the Honda Civic that run on natural gas or the Smart ForTwo (at almost one quarter of the impact of the Prius) may be better overall (see Table 3.2). Again, we just don't know because no one has access to the numbers to do a real evaluation (and none of the car companies is offering up their data).

TABLE 3.2

Dust to Dust Cost per Mile							
	Smart ForTwo	*Corolla*	*Camry*	*Civic*	*Prius*	*Civic Hybrid*	*Hummer H2*
2008	.583	.748	2.167	2.867	2.191	2.943	3.621
2004	NA	.732	1.954	2.867	3.25	3.25	3.027

To be sure, driving any car less, carpooling, and driving more efficiently are better solutions. On the whole, driving lighter cars made closer to home is also better. But how many customers do you know who are able to source the data for where a car is manufactured—especially when most cars are made from subassemblies from all over the world?

The point is that we don't have this data yet and even if companies have the data, they aren't compelled to share it with us. And the data we use, and our assumptions about it, make all of the difference in our evaluations. That means we can't make informed decisions.

Using an LCA as a tool in the design and development process requires that choices be made as to the types and amounts of materials that will be used in the product. LCA tools can be helpful in choosing among materials in the early phases of a project, but cannot be fully applied to assess the environmental impact of the finished product until it is fully specified and designed. **This makes it difficult to use during the conceptual phases of development.**

Shortcuts can sometimes be used to speed the process by concentrating on the biggest impacts, but this is hardly fool-proof since it's based on assumptions that may be wrong. By looking at the largest components— and often the costliest materials—and involving experts with material impact experience, "rough guesses" of a concept's impact can often be derived that are "good enough" for the purposes of concept selection. This is called *matrix LCA* (as opposed to *full-scale LCA*). Ecological footprint calculators, for example, work along the same principles. Proxy LCA uses average weightings of material impact with simplified accounting and is the most common type of LCA to be used by designers during conceptualization and prototyping.[8]

Despite some of the drawbacks, LCA tools are improving constantly, and they are still the best way to assess environmental and resource impact with any degree of accuracy. LCA tools focus on environmental impact, but don't take into account the social impact of a product. We now will look at tools that evaluate social impact as well.

---

[8] www.pre.nl/download/EI99_Manual.pdf

Social Return on Investment

Social Return on Investment (SROI) strategies attempt to measure social-economic impacts like LCAs measure environmental impacts (see Figure 3.14). This is a very new approach, and there are no still no standards or agreement about how best to do this. It also suffers the challenge of *which* social issues to measure, since any list will likely be different for each individual or group.

Organizationally-developed Criteria

FIGURE 3.14
Measuring Social Return on Investment.

Strengths: Describes an approach to valuing social issues within a financial framework. Open to interpreting approach for specific conditions.

Weaknesses: Controversy over defining of social value in financial terms. No set template for developers to follow. Subjective.

However, progress is being made. Socially Responsible Investment Funds (SRIs), for example, use proprietary metrics to measure the social performance of the companies they invest in and "screens" to bar investments unmatched with their values. All of them publish the list of companies they invest in and some publish their metrics, but there is little agreement in approach or values.

Generating a Social Return on Investment (SROI) assessment is not a trivial undertaking, because it requires the translation of social values into some type of monetary return. Considerable controversy surrounds the notion that values such as tolerance and anti-discrimination can be translated into monetary terms easily or at all. Some argue that it would be wrong to even attempt to monetize core human values, as they fear that inappropriate trade-offs will result. For example, is a little slavery all right if the bottom line improves?

There are several different approaches to calculating SROI, but one of the most well known is the Social Impact Assessment used as part of the Haas Global Social Venture Competition, orchestrated by Haas Business School at UC Berkeley. This structure has evolved into a framework, templates, and cases that describe how some social ventures are attempting to value the social component of their business' impact. SROI can be calculated for any organization, non-profit and for-profit alike, at **www.gsvc.org**.

To start with, it's important to establish and describe the impact value chain across the social issue spectrum. These track the inputs (benefits) of the solutions, the activities that create the impact, the outputs (value being generated in social terms), and the overall outcome of the organization's impact. To be sure, it's not possible to calculate all social impacts, and it can be controversial how an organization makes its assumptions and calculations. More so, it may not be appropriate to even try to quantify social impact for some cases (such as trying to justify freedom over slavery—there shouldn't need to be a quantification).

SROI of Reveal Sustainability Labeling Initiative

The Reveal system is the result of several years of my work and my thesis at business school. It is a rating and labeling system for use with products and services in order to help customers make more informed choices at the points of decisions and purchase. More on the solution itself in Chapter 17. When I was preparing the business plan, in order to describe the SROI, I used the process outlined by the Global Social Venture Competition (which we entered) and had to find a way to reasonably measure the impacts such buying decisions had in economic terms.

SROI of Reveal Sustainability Labeling Initiative (continued)

To be conservative, I chose to look at only three products: LED light bulbs, compact fluorescent light bulbs, and low-consumption power adapters. By no means would this represent the total buying power of consumers making choices for more sustainable products, but it was a simple example.

I measured the economic impact for reducing only carbon dioxide and sulfur dioxide in the environment (see Figure 3.15). For sure, there would be many other positive effects for these products. I valued these at the current price in the Chicago Climate Exchange (CCX) trading markets and made as conservative assumptions as I could about customer adoption (5 percent adoption in the beginning, growing to 80 percent in 10 years, from only the most environmentally-oriented 17 percent of the U.S. and EU populations). I also assumed that customers would only replace three CFL bulbs, one LED bulb, and one low-consumption power adapter per year. These simple purchases would result in over $100B of benefits over 10 years (see Figure 3.16). Now, imagine what would happen if more than merely one percent of the U.S. and EU populations replaced three bulbs and one power adapter every year. Or imagine the gains that all of their other purchases would create if they made more sustainable choices.

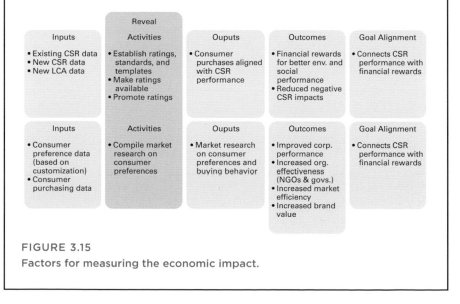

FIGURE 3.15
Factors for measuring the economic impact.

SROI of Reveal Sustainability Labeling Initiative (continued)

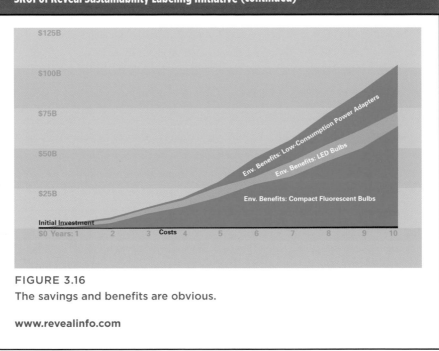

FIGURE 3.16

The savings and benefits are obvious.

www.revealinfo.com

The Natural Step™

The Natural Step is a framework (see Figure 3.17) promoted by an international organization that proposes four fundamental system conditions to help stabilize the global biosphere. This approach is general, and the ecological and economic benefits of the Natural Step are often difficult to measure (which isn't so different than many of the other frameworks).

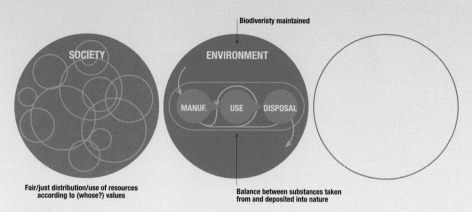

FIGURE 3.17
The Natural Step framework.

Strengths: A development and business-positive approach seeking to help organizations make better choices.

Weaknesses: No formal techniques to use for either evaluation or development. Subjective.

The four system conditions include the following:

- **System Condition #1:** Substances from the earth's crust must not systematically increase in the biosphere. Fossil fuels, metals, and other minerals must not be extracted at a faster rate than their redeposit and regeneration in the earth's crust.

- **System Condition #2:** Substances produced by society must not systematically increase in nature. Substances must not be produced faster than they can be broken down and be reintegrated into the cycles of nature or be deposited in the earth's crust.

- **System Condition #3:** The physical basis for the productivity and diversity of nature must not systematically deteriorate. Productive surfaces of nature must not be diminished in quality or quantity, and we must not harvest more from nature than can be recreated or renewed.

- **System Condition #4:** There needs to be fair and efficient use of resources with respect to meeting human needs. Basic human needs must be met with the most resource efficient methods possible, including equitable resource distribution.

The Natural Step advocates systems thinking. Like Natural Capitalism and Proxy LCAs, it tries to focus on the sources of the greatest impacts in order to generate progress without getting bogged down in disorienting details. This is best displayed by The Natural Step Resource Funnel (see Figure 3.18).

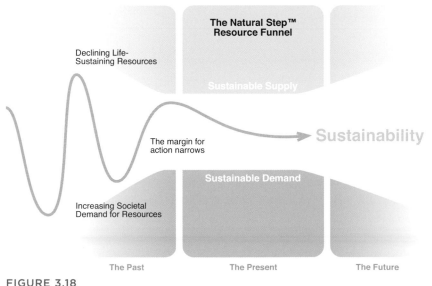

FIGURE 3.18
The Natural Step Resource Funnel.

The funnel presents the immediacy of the global ecological crisis in a funnel-shaped diagram, and shows how important it is for us to acknowledge the crisis and its impacts. The funnel shows how the decline in living systems is in conflict with the increasing demand for products. As the funnel narrows, there is less and less margin for action. However, it demonstrates that forward-looking organizations that act on this understanding will be better suited to balance sustainable demand with sustainable supply. Businesses can be shown that it is economically advantageous to make change now, before future pressures make it necessary—because a forward planning company (moving through the center of the tunnel) is more profitable than a reactive, defensive company. The ultimate goal of Natural Step's perspective on business is the restoration of living systems.

Total Beauty

Created by Edwin Datcheski to redefine the concept of what is "beautiful," Total Beauty is a quantitative framework that offers a point system to calculate total impact of products and services in environmental terms (see Figure 3.19).

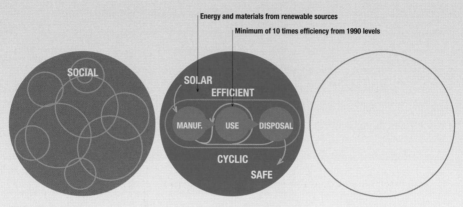

FIGURE 3.19
The Total Beauty framework.

Strengths: Easy process for designers to use. Can be used in the design process, as well as to evaluate existing products. Provides a fast assessment that is (relatively) easy to use. Favors specific materials, including natural and organic materials, by embodied energy, over technical materials (which biases against products, such as electronics).

Weaknesses: Subjective. Incomplete (not detailed). Social issues are left incomplete. Complex products naturally have lower scores than simple products (which may not be an accurate assessment, considering a product's performance or functions).

Consistent with this book's approach to design, Edwin Datschefski defines *beauty* as being inclusive of sustainability. To him, unsustainable products can't possibly be beautiful. He defines five unique categories of criteria to encompass sustainability concerns, including the following:

- **Cyclic** refers to solutions that are manufactured in processes that close the loop between resources used and wastes generated. This includes recycling, composting, using organic materials, and stewardship over sustainable sourcing.

- **Solar** refers to solutions that use renewable sources of materials and energy during the manufacturing and use phases (such as wind, solar, geothermal, and existing hydro) and must be produced according to the other criteria (for example, Cyclic, Safe, and Social). Likewise, materials that are grown sustainably and are easily and realistically replenished are preferable to those that aren't. The three categories of energy this framework promotes are muscle power (both human and animal), hydrogen and electricity, and photons (both photosynthesis and solar panels). Each of these represents a form of power ultimately derived from the sun. Note: Hydroelectric power is listed as "renewable" but remains controversial for newly proposed and many existing power plants for other environmental and social reasons.

- **Safe** is self-explanatory and extends not only to product safety but also to toxicity of materials, concentration, and location. In fact, all releases from the manufacturing, use, or disposal of products to the air, land, water, or space should be safe for people, animals, plants, and the environment as a whole. This concept reflects the understanding, popular in other frameworks, that all outputs of one process are either the inputs to other processes or at least affect these inputs. For example, heat isn't generally unsafe. However, when it's concentrated as run-off from a power plant cooling system dumped into a river, it can easily kill all aquatic life in the vicinity (plants, animals, bacteria, etc.).

- **Efficient** refers to the goal of reducing energy and material use (including water) in manufacturing to 90 percent of average levels in 1990. Admittedly, 1990 is an arbitrary date, but it's an acceptable goal (one has to start somewhere). Is a 90 percent reduction realistic and enough of a change? Eventually, probably not. However, it's a challenging goal at present and when it becomes less of a challenge, there will be the opportunity to change this goal again.

Quantitative by nature, this criteria is one of the easiest to measure, but can be one of the most difficult to achieve—especially for well-developed, simple solutions such as flower pots or woolen scarves. The efficiency of solutions that have been in use for decades or centuries may have been improved so much over time that it is difficult today to achieve the gains required by this criteria. At some point, it becomes less and less possible to eke out more efficiency as solutions encounter physical limits. However, we're far from this point currently.

Efficiency in this context also includes (and is measured by) parameters such as utility, durability, upgradeablity, reparability, complementary components, long-term thinking, increased efficiency, increased utility, dematerialization, multifunctionality, and locality.

- **Social** refers to the desire for solutions to support basic human rights and "natural justice." Datschefski doesn't elaborate and go into as much detail about the myriad social issues described in the previous chapter (and who can blame him), but this is a container for all of those issues that pertain to people (not animals). It can be problematic, however, as it is heavily reliant on specific cultural values. What is acceptable in some cultures may not be in others. In addition, there are many degrees of compliance to social issues. For example, to some, animal testing may not be acceptable, but eating animal products may. To others, neither may be acceptable. Reading Datschefski's criteria literally, none of these distinctions applies.

Because there are five categories described, Datschefski's system is easy to remember. However, many designers have problems understanding the scope of *Cyclic* and *Solar* because they encompass what often seems like unrelated criteria. It's also not as obvious how to measure or approach these first two goals as it is for *Efficient* and *Safe.*

Datchefski introduces a scoring system to assess the "beauty" of products against each of his five criteria. This system is more easily applied to existing products than it is to creating new solutions, although it might provide a starting point in thinking about improvements or replacements to existing products.

To score a product in the Cyclic category, Datschefski's equations calculate the amount of recycled materials used in manufacturing plus the amount of product material that is recycled when the product is discarded, and divides this by 2:

Cyclic (%) = (% recycled content + % material recycled) / 2

To score a product in the Efficient category, the equation is a little more complex:

Efficient (%) = 100 $(1 - \frac{1}{N} \sum_1^N \left[\frac{today}{1990} \right])$

Solar (%) = % of total energy from renewable sources[9]

Safe (%) = % non-toxic lifetime releases of all outputs (to air, water, waste, etc.)

Apparently, **Social** isn't scored. Instead, products and services that don't rate well socially shouldn't be considered further.

This scoring system includes *ugly points* for particularly bad performance criteria, materials use, energy inefficiency, or social ills. This is the simplest and easiest part of the system to calculate. As with EIO-LCA, it is based on rough assumptions and averages, not on actual data from measurements. Examples of ugly points for different materials include the following:

- −1 for every kilogram of bioplastics, ceramics, asphalt, concrete, wood, stone, and brick

- −5 for every kilogram of food, glass, most plastics, paper, rubber, steel, textiles, clothes, furniture, gas, diesel, carpet,[10] etc.

---

9   This includes embodied energy. Refer to the LCA framework to appreciate how this must be calculated throughout the life cycle. Renewable sources include solar, wind, muscle, photosynthetic, geothermal, hydro, and wave power.

10 However, not all carpet is the same, so these points are crude numbers based on averages and assumptions. Carpet from Interface, for example, is distinguished in every way from carpet produced by the rest of the industry (**www.interfacesustainability.com**).

- −15 for every kilogram of aluminum, light bulbs, paint, plastics like polystyrene and polycarbonate, stainless steel, and electronic assemblies

- −50 for every kilogram of gold, lead, brass, nickel, copper, chromium, chromed steel, cadmium, zinc, and batteries of any kind

Also, it's possible for the same element to score differently—sometimes drastically—in different categories. For example, even if a material is grown organically and harvested by equipment that is powered by renewable sources (scoring highly in the Cyclic and Solar categories), it can still be harvested by slaves (scoring low in the Social category) or be an unsafe material (scoring low in the Safe category).

Likewise, although Datschefski's criteria favor solutions that are biomimetic or "natural," they may still be socially acceptable.

Total Beauty can be scored relatively or absolutely, meaning the points awarded in scores can be calculated from absolute quantities for a product or assessed in relation to competing alternatives (such as scoring and adding plusses and minuses in performance categories instead of numbers to give a relative comparison between alternatives). Either way, Total Beauty, like most of the other frameworks, isn't valuable for its accuracy so much as the general impression the assessment gives of a product or service and its impact in regards to specific criteria. Often, this is what designers need to know most, especially during the concept and prototype phases.

An example might help you better understand this system.

Eva Solo Flowerpot

Designed by Eva Solo Denmark, this self-watering flowerpot has won numerous design awards, including a 2005 International Design Excellence Awards (IDEA) Silver Medal. The pot helps reassure users that their houseplants won't die from over- or under-watering, as the flowerpot always delivers the correct amount of water to the plant via a wick (see Figure 3.20). The design is considered to be aesthetic and functional, but is it sustainable?

Eva Solo Flowerpot (continued)

FIGURE 3.20
A self-watering flowerpot—what an innovation!

The entire product consists of three parts: a ceramic flowerpot, a nylon wick, and a glass bucket that functions as a base. The product is sold primarily through online retailers for \$35–\$40, and is produced in two sizes with various colors. The flowerpot itself is made of advanced ceramics. A 9.5" nylon wick is laced through the bottom holes of the pot, and dangles into the water bucket underneath, allowing the plant to absorb water as necessary, mimicking the plant's own roots. The glass bucket holds roughly one week's worth of water for a typical houseplant and is translucent to easily see if the water level is low.

Cyclic: 61.5% (60% for the ceramic + 61% for the glass + 35% for the nylon wick + 90% organic, divided by 4)

Solar: 28.75% (5% for the ceramic + 5% for the glass + 10% for the nylon wick + 95% organic)

Safe: 57.5% (45% for the ceramic + 30% for the glass + 60% for the nylon wick + 95% organic)

Efficient: 50%

Social: 87% (based on issues with nylon production)

Ugly points: 2.25

Eva Solo Flowerpot (continued)

Note: One of the issues with the Total Beauty framework is that products made from traditional materials and processes suffer because recent efficiencies are much less likely compared to what has been achieved over hundreds of years. Ceramic and glass-making are old technologies whose efficiencies have been reasonably achieved for a long time—certainly predating 1990 by many decades. In other words, they may be as efficient as they can be made. This penalizes older materials and processes.

Though the product scores relatively well already, the team that evaluated it was still able to identify some potential improvements:

- Sourcing the energy used to make the components from renewable sources

- Using traditional ceramic (such as clay refractory) instead of the advanced ceramic

- Using recycled materials as much as possible in the manufacture of the three parts

- Using an organic cotton wick instead of nylon

- Manufacturing in several locations to reduce transportation

Evaluation by Eunice Barnett, Lindsay Clark, Stephen Lamm, Hillary Meredith, and Daniel Winokur, Presidio School of Management, 2008.

www.evasolo.com

Sustainability Helix

The Sustainability Helix[11] is the result of collaboration between Natural
Capital Solutions[12] and students from the Presidio School of Management.[13]
It is a framework for evaluating overall organizational commitment and
progress in sustainability. Unlike many frameworks, it is decidedly
business-positive, and it describes a clear path from wherever an
organization might score initially toward greater sustainable development,
processes, and strategy (see Figure 3.21).

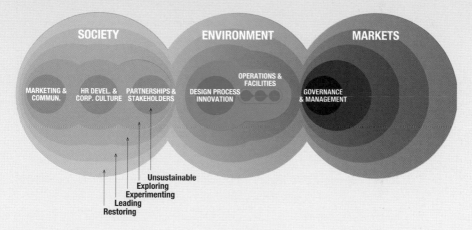

FIGURE 3.21
The Sustainability Helix framework covers all domains of sustainability.

Strengths: Business-positive and not moralizing. Neutral assessment
of position along a spectrum. Promotes and contextualizes restoration,
not merely sustainability. Integrates business functions throughout an
organization and promotes involvement and cooperation of these over time.

Weaknesses: Lacks metrics for measurement. Subjective.

---

[11] Chapter 10 of *The Natural Advantage of Nations*, (Earthscan 2005)

[12] www.natcapsolutions.org/, 3/20/2008

[13] www.presidiomba.org/, 3/20/2008

The Sustainability Helix describes five stages through which an organization progresses from least to most sustainable performance:

- Stage 0: Unsustainable
- Stage 1: Exploration
- Stage 2: Experimentation
- Stage 3: Leadership
- Stage 4: Restoration

At each of these stages, the framework describes measures to assess how well organizations are doing, desired outcomes that describe what the organization should aspire to achieve at that stage, and strategies for moving to the next stage. By including the lowest level, unsustainable, the helix provides all organizations with a place to position themselves on the helix, even those that have not yet considered sustainability.

The helix tracks progress in six categories of an organization's operations and management (see Figure 3.22), allowing the organization to coordinate strategy, progress, and tactics across all functions in order to maximize the benefits and optimize the effectiveness of any given strategy or tactic. The six categories include the following:

- Governance and Management
- Operations and Facilities
- Design and Process Innovation
- Human Resources and Corporate Culture
- Marketing and Communications
- Partnerships and Stakeholder Engagement

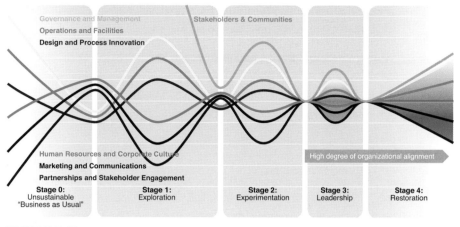

Governance and Management
Operations and Facilities
Design and Process Innovation

Stakeholders & Communities

Human Resources and Corporate Culture
Marketing and Communications
Partnerships and Stakeholder Engagement

High degree of organizational alignment

Stage 0:	**Stage 1:**	**Stage 2:**	**Stage 3:**	**Stage 4:**
Unsustainable	Exploration	Experimentation	Leadership	Restoration
"Business as Usual"				

FIGURE 3.22

The Sustainability Helix (www.cmcusa.org/initiatives/helixoverview.cfm).

The third category, Design/Process Innovation, specifically addresses product and service development, but all of the categories relate to development, to some extent. Coordinating across all of them will be more powerful than confining development to its silo. Although the Sustainability Helix is more a strategic tool for corporate leadership, its framework is something all aspects of an organization, including product/service development, should be aware of, understand, and invoke in creating new solutions.

> Although the Sustainability Helix is more a strategic tool for corporate leadership, its framework is something all aspects of an organization, including product/service development, should be aware of, understand, and invoke in creating new solutions.

The Helix describes an exhaustive list of strategies throughout an organization, including operations strategies such as the following.

Governance and Management Strategies:

- Ethics Strategy

- Project Development Systems

- Risk and Shareholder Value: Insurance Strategy

- Unsustainable Life Cycle Risks Strategy

- Shareholder Value Strategy

- Legal: Regulatory Strategy

- Legal: Intellectual Property Strategy

Some Operations Strategies:

- Manufacturing and Production Procurement Strategy

- Sustainable Manufacturing Strategy

- Zero Waste Strategy

- Product Stewardship Strategy

- Facilities Resource Procurement Strategy

- Facilities Location Strategy

Design and Process Innovation Strategies:

- Whole-Systems Design Strategy

- Product/Service Design Strategy

- Process Design Strategy

- Testing Strategy

Human Resources Strategies:

- Optimal Utilization Strategy

- Internal Training Strategy

- Incentive Systems

- Sustainability/Risk-Taking Culture Strategy

- Cultural Alignment Strategy

- Healthy Workforce Strategy

- Diversity Strategy

Marketing and Communications Strategies:

- Sustainable Marketing Strategy

- Sustainability Education Strategy

- Internal Communications Strategy

- Public Relations Strategy

- Branding Strategy

Partnerships and Stakeholder Strategies:

- Research/Innovation/Risk Partnerships

- Regulatory Partnerships

- Human Capital Strategy (External)

- Value Chain—Customer Strategy

- Value Chain—Supplier Strategy

Most relevant to product and service developers are the list of strategies specific to Design and Process Innovation, including Whole System Design, Product/Service Design, Process Design, and Testing. These relate directly to the actions that designers can take today, described in detail next. Whole Systems Design acknowledges the need for organizations to address the full life cycle of their products and services and engage partners, customers, and other stakeholders in order to research, innovate, and reduce risk. Product/Service Design strategies seek to have developers readdress the value chain in solution development and fulfillment under the principles such as Natural Capitalism, in order to take the actions described next, resulting in more sustainable solutions. Process Design strategies seek to help developers better understand how their processes can create incentives and realize more sustainable solutions. Lastly, Testing strategies help developers ensure that their solutions reduce impacts and are actually more sustainable.

As with the other frameworks, the detail the Helix describes requires deeper exploration, Fortunately, all of the information about the framework is available online.

Other Frameworks

These are, by no means, the only frameworks in use. There are several more and, undoubtedly, there will be new ones soon. Some of the others are focused on government compliance, such as the GRI (Global Reporting Initiative—see Figure 3.23), ISO 14000, ISO 16000, and SA 8000 standards. All are relevant and form a piece of the sustainability mosaic, but organizations will need to review and engage several frameworks in order to become more sustainable.

FIGURE 3.23
GRI Categories and Aspects diagram by Covive: www.covive.com/gri/.
GRI Framework by the Global Reporting Initiative.

Another framework, designed for rating materials in the building industry is Pharos (see Figure 3.24). It's still new and in development, but it offers designers in the building domain a guide to making material selection based on environmental and social impacts.

FIGURE 3.24
Pharos, a framework for the building industry.

Putting Them All Together

The good news is that, despite how different many of the frameworks seem—all of them attempt to achieve similar ends, and there is a lot more overlap and similarity between them than differences. Taken as a whole, each framework can be layered on top of the rest to form a more complete summary that presents a coherent picture of what sustainability means to an organization and to a design process (see Figure 3.25).

In addition, even the "quantitative" frameworks run the risk of being qualitative in the way they are measured and assessed. This is because there are few standards for what factors are covered, how, and what measures to use.

You can use this summary framework as a starting point for addressing sustainability in your organization, your development process, and the sustainability impacts of the solutions you create (see Figure 3.26).

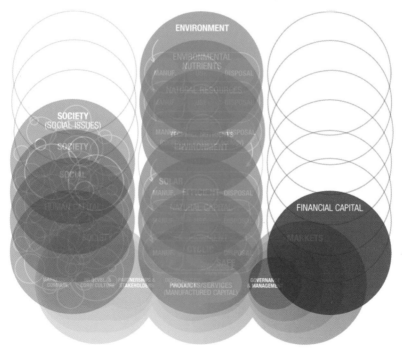

FIGURE 3.25
The juxtaposition of all the frameworks.

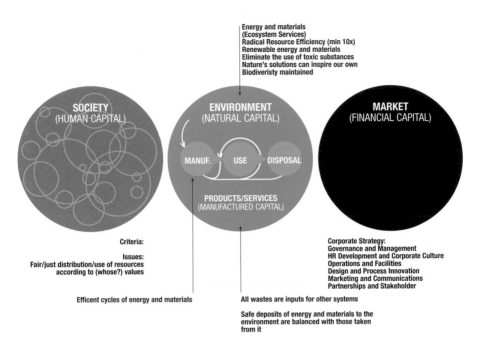

Energy and materials
(Ecosystem Services)
Radical Resource Efficiency (min 10x)
Renewable energy and materials
Eliminate the use of toxic substances
Nature's solutions can inspire our own
Biodiveristy maintained

SOCIETY
(HUMAN CAPITAL)

ENVIRONMENT
(NATURAL CAPITAL)

MARKET
(FINANCIAL CAPITAL)

MANUF.   USE   DISPOSAL

PRODUCTS/SERVICES
(MANUFACTURED CAPITAL)

Criteria:

Issues:
Fair/just distribution/use of resources
according to (whose?) values

Corporate Strategy:
Governance and Management
HR Development and Corporate Culture
Operations and Facilities
Design and Process Innovation
Marketing and Communications
Partnerships and Stakeholder

Efficent cycles of energy and materials

All wastes are inputs for other systems

Safe deposits of energy and materials to the
environment are balanced with those taken
from it

FIGURE 3.26
Building a summary framework.

Reduce

Perhaps the most important thing that designers and developers can influence is the reduction of materials and energy used in the products, services, and events that enhance our lives. The famous phrase less is more was never more appropriate or important than in the context of environmental, social, and financial sustainability.

While some environmental organizations focus on corporate practices or reporting, the biggest impact that organizations make is in the materials and energy associated with their products and services. This should be the primary measure of how committed a company is in improving the sustainability (across all categories) of their offerings.

CHAPTER 4

Design for Use

O ne of the most important and seemingly simple design principles developers can employ is to make sure that the things they design are usable. Not only does this increase the likelihood that your customers will prefer to buy your solutions, but it also increases the likelihood that, once they do, these solutions will actually be used, instead of discarded for something else that might work better. I'm sure you've been in this situation yourself, where you purchased something only to find out it didn't fit your needs and you had to go find a substitute. Two or more solutions created to solve one problem is *not* a strategy for sustainability. Therefore, the more effective your solution is, for those for whom it is intended, the more likely that it will be used.

There are a few principles that fall into this category that can help focus designers and developers on creating more successful solutions. For example, usability, clarity, accessibility, and meaning each contributes to more usable solutions that stand a better chance of being used over a longer period of time.

> **Two or more solutions created to solve one problem is *not* a strategy for sustainability.**

Usability

The idea that solutions should be easy to use is hardly new, but frankly, it still doesn't always result in usable products and services. This is especially true with technological solutions, although even common objects around the house could be improved. For example, OXO's lines of Good Grips™ products are exceedingly well designed—make that redesigned—products that seem obvious once you pick them up and use them. Though they were designed for seniors with difficulty gripping traditional utensils, they've found a much wider audience because their grips are easier to hold and use. Many of the items introduce small improvements that make the products more useful (such as measuring cups that are easier to read accurately). However, improvement potential lurks within everything.

A recent Accenture study reported that 95 percent of product returns had nothing to do with the product's functionality. (In these cases, the products were all functioning as designed.) While 27 percent of returns reflected buyer's remorse, the rest were due to either products that didn't fit perceived needs or products that users couldn't use effectively.[1] The 11–20 percent of products that were returned in the consumer electronics industry alone represented $13.8B worth of merchandise that could not be resold as new. This dollar amount not only represents an exorbitant financial cost to companies but also a staggering material and energy cost to society and the environment. So obviously, reducing usability issues could have a drastic impact on product and service effectiveness.

Much is known about how to make solutions more usable for people so I won't go into detail about that in this book. (Appendix B at the end, however, lists several great books on the subject.) However, here is a quick overview of some of the most popular and important usability principles.

Principles of Usability:

- **Design for People.** Research and understand who it is that you're developing solutions for, whether products, services, or events, and understand their needs, desires, and contexts.

- **Feedback.** Make sure that users can see the system's status at all times, as well as where they are in the system or process, what they're doing, and what is available to them.

- **Familiarity.** The solution should reflect users' language, customs, and understanding, rather than technical conventions.

- **Forgiveness.** Systems should help users understand and recover from mistakes easily, such as the near-ubiquitous Undo and Redo commands in computer software.

- **Consistency.** Wherever possible and appropriate, solutions should be designed with consistent language, frameworks, contexts, and communications.

---

[1] www.theinquirer.net/gb/inquirer/news/2008/06/03/per-cent-returned-electronics

- **Efficiency.** Superior systems should allow both novices and experts to work at their different paces, providing tools for each to work at their level of proficiency.

- **Direct Manipulation.** System elements should be directly active, rather than through proxies, allowing users to better use and understand the system and how to use it.

- **Natural Cognitive Structures.** The perspective presented of the system, its parts, and its organizational structure should be reasonably accurate, reflect the activities within the system, promote understanding about its use, and reduce the demand for users to memorize actions and behaviors.

Simplicity Versus Clarity

One method that designers, developers, and (especially) marketers often use to increase usability is to make things "simple." Often, the approach is simply to take away features and performance criteria to leave only a few. This isn't a terrible strategy—that is, if you know which criteria are the most important to keep in terms of the solution's intended performance and in terms of what customers actually need. As simple as this sounds, most organizations are terrible at identifying what outcomes customers actually need. Instead they try to separate product features into identified layers of "consumer" (often lacking critical features), expert or professional (often packing features without improving how they're presented), and enterprise (often ignoring usability altogether under the assumption that users will want to either configure the interface themselves or have no problem finding the signals they want among the noise).

Most businesses think that simplicity is the only way to make solutions more usable, but this is neither foolproof nor the best method. Often, simply deleting features renders a product or service more ineffective. Complexity, in itself, isn't bad. In fact, it's often critical to both understanding and usability.

Maps, for example, can be extremely complex. Yet if we were to simply reduce the data on them, they would cease to work well because the context needed for understanding them would be altered, often eliminating the possibility they could be used for more than one purpose. Instead, the critical ingredient doesn't involve deleting the data, but rather making its organization and presentation clearer for people. Simplicity can be a winning strategy sometimes, but clarity is always required, no matter what the level of complexity is.

> **Most businesses think that simplicity is the only way to make solutions more usable, but this is neither fool proof nor the best method.**

A simple map, like a simple life, isn't terribly representative of most people's experiences. When solutions are simplified, they usually become less effective because they no longer apply to the complex experiences we encounter. Instead, rendering the complex clear reflects the reality of most systems, solutions, and experiences without sacrificing understanding or usability. To do this, clarity relies on the prioritization of cognitive models and features that are most important, while downplaying those that are less critical. This can be done through careful arrangement of elements in the visual, auditory, temporal, and other sensorial dimensions.

Accessibility

Making sure that a solution is usable for a wide variety of people, with different abilities, understandings, and capabilities, ensures that the solution can be used as much as possible. This isn't just a matter of developing for different physical and mental abilities (though that's important, especially from a social justice perspective). Accessibility also refers to the different modes we all operate under at different times. For example, sometimes we search, sometimes we browse. Sometimes, we're lightning-focused, and other times we prefer to meander. There are contexts where a visual interface is more easily used, and other times when a verbal, auditory, or even textural interface might be more appropriate.

> **Designing for accessibility requires us to explore a full range of uses and modes in order to develop solutions that address the widest array of people possible.**

Using the example of a map again, a long list of directions might be best rendered on a map instead of a text list since there are few visual cues in the list to help drivers know where and when to turn. A page of text, in either a list or story, doesn't make it as easy to pick out the street names, map directions, and so on, as in a graphic map. However, given driving requirements, a geographically-appropriate voice system might be even better than a map since the instructions can be spoken clearly precisely when they are needed, allowing drivers to keep their eyes on the road and not on a map or list. Of course, the most effective systems combine several types of renderings to increase their accessibility and usability further.

Designing for accessibility requires us to explore a full range of uses and modes in order to develop solutions that address the widest array of people possible. This awareness not only makes our solutions usable for a wider range of people, but it also makes them more useful across a wider range of contexts.

A selection of accessibility issues includes the following areas:

- Hearing impairment

- Speech impairment

- Vision impairment

- Dexterity impairment

- Color blindness (several types)

- Loss of balance or mobility due to age or illness

- Searching contexts versus browsing contexts

- Entertainment versus information contexts

Meaning

Meaning has been a growing point of discussion in the design world over the past four years. Even in the business world, *meaning* is increasingly addressed by strategists, entrepreneurs, and investors, especially in the sustainability and social venture markets. Authors such as Guy Kawasaki[2] regularly extol the need for developers and organizations to make more meaningful offerings (see Figure 4.1).

Typically, designers aim to engage customers on an emotional level, and we find that the discussion of the virtues and dangers of emotional engagement usually includes examples from advertising. However, developers of all types rely on instinct to engage customers at this level, and they lack a coherent, reliable framework for engaging customers more deeply and consistently in product and service development. Whether the target emotions are positive (love, admiration, joy, excitement, etc.) or negative (fear, terror, anxiety, etc.) doesn't change the process. Emotions are powerful mechanisms for customers to connect and build relationships with (or revulsions to) products, services, events, brands, or organizations.

> **Emotions are powerful mechanisms for customers to connect and build relationships with (or revulsions to) products, services, events, brands, or organizations.**

The economist, Tibor Scitovsky, has developed a theory of joy, explained in his book, *The Joyless Economy.* He observes that more money and material possessions often result in more dissatisfaction in people's lives. The book describes how comfort and pleasure are opposed, and how both specialization and mass production contribute to an economy of ever-greater stimulus (through novelty) and often less satisfaction. Simply providing more and more novel solutions doesn't result in a higher quality of life. (For example, just look at the proliferation, spikes, and downfalls of

[2] www.makingmeaning.org
    blog.guykawasaki.com/files/Art.pdf

most new kitchen appliances, such as the Hot Dogger and automatic bread maker.) Emotions are quick phenomena that are powerful and motivating, but are not sustainable over medium or long periods of time. However, when we set expectations for continued levels of high emotions, we create the conditions for even higher disappointment and dissatisfaction.

Emotions aren't even the most powerful or deepest level of connection that customers make with products, services, events, environments, or brands. Values and meaning run even deeper than emotions and require even more careful consideration. All of these, however, are now described in a framework for designers to use in the development process.

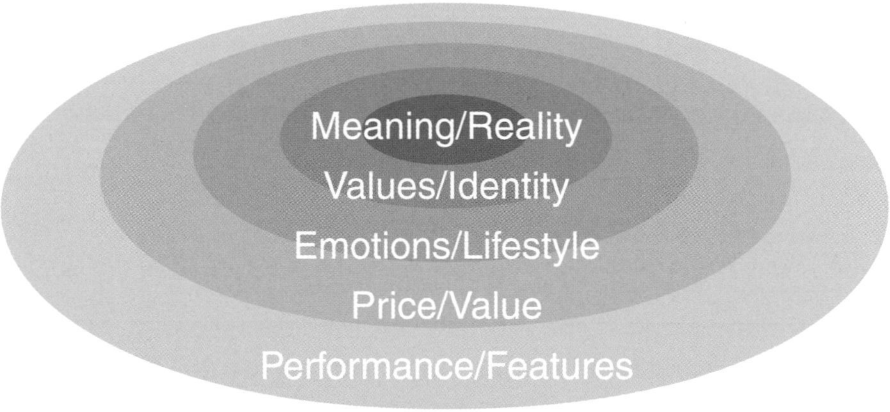

FIGURE 4.1
The levels of meaning.

> **Values and meaning run even deeper than emotions and require even more careful consideration.**

Meaning is the deepest level at which people (audiences, users, customers, participants, etc.) engage with a product, service, or event. It is the most important aspect of the experience created between people and objects or between people and others. It represents the deepest level of five levels that describe how we assess (often unconsciously) how these experiences fit into our lives and how they are important to us. In addition, it's used as an evaluation criteria for choosing among the myriad alternatives and options available to us.

The five levels of significance include the following areas (see Figure 4.1):

- Meaning/Reality

- Identity/Values

- Emotions/Lifestyle

- Price/Value

- Performance/Features

The 15 core meanings are represented by the following attributes:

- Accomplishment

- Beauty

- Creation

- Community

- Duty

- Enlightenment

- Freedom

- Harmony

- Justice

- Oneness

- Redemption

- Security

- Truth

- Validation

- Wonder

While it hasn't been proven, there is circumstantial evidence that people who are engaged by a product at the deeper levels of significance (such as values and meaning) are less likely to consume products and services simply to fill some perceived void or dissatisfaction. If you think about the people in your life who consciously choose *not* to consume at "normal" levels or make it a point to live "lightly," they are often those who are most satisfied in terms of finding meaning in their lives and living according to their values. Again, I haven't seen definitive proof of this yet, but my instinct says there is a pattern here.

If this is indeed the case, then one path toward more sustainable solutions is in more meaningful ones. If we consume less when our meaning and value needs are satisfied, then meaningful experiences and solutions can lead to more sustainable ones. Think about the things in your life that you cherish. These aren't things that you give up or replace lightly. Chances are that they're not interchangeable within the context of fashion and trends and that you don't feel the need to discard them easily.

> **While it hasn't been proven, there is circumstantial evidence that people who are engaged by a product at the deeper levels of significance (such as values and meaning) are less likely to consume products and services simply to fill some perceived void or dissatisfaction.**

Those who engage the world in meaningful ways don't look to products and services so much to satisfy their core meanings. For example, when your need for oneness and community is fulfilled, you will certainly buy clothes that reflect these meanings for you but you will likely not need to buy so many clothes to accomplish the same meaningful sensation. In addition, because these clothes connect to your values and meanings so deeply, you won't be as willing to discard and replace them with others simply because styles have changed. The result is that you may not only buy fewer clothes and keep them longer, but you will feel a deeper connection to them and feel more accurately represented by them.

Those who engage the world in meaningful ways don't look to products and services so much to satisfy their core meanings.

For sure, the items, context, and foci of these deep meanings change individually and most people won't reduce their consumption simply because they can. Some may find connection in clothes or food or cars or their home, while others might find it in their work, their hobbies, or the time they spend with people. We can't prioritize all 15 core meanings in our lives. At best, we can prioritize between two and four. In addition, we all express these meanings differently, depending on our personal and shared values. However, these core meanings are universal throughout all cultures so we have a point of departure where we can address meaning from a common framework. This is why design research is so critical during the design process—it serves to uncover the elements that can lead us to better understand our customers at these levels.

CHAPTER 5

Dematerialization

A nything developers can do to reduce the amount of materials and energy in a solution will reduce—sometimes dramatically—the impact it has on resources and the environment. Systematic reduction of toxic materials, careful engineering to reduce product size and weight, and reduction of waste and energy in the manufacturing process (where 80–90 percent of a product's impact often occurs) can make a tremendous improvement in a solution's sustainability. These gains compound as well. For example, a lighter product improves fuel efficiency and reduces waste exhaust everywhere it travels (if it's shipped, trucked, or driven). Some solutions actually dematerialize other products (see the "iPhone" sidebar). What's helpful about this approach is that, most of the time, reducing the materials and energy required to make and maintain products and services often reduces their costs. Therefore, this design technique directly benefits businesses and the environment, which makes for an easy sell to even the most narrow-minded businesspeople.

Design for Efficiency

Perhaps the most important strategy for improving sustainability is to increase the efficiency of material and energy use through all parts of a product's life cycle. With that in mind, eco-efficiency proposes that we address and invent new processes, arrangements, and implementations in order to reduce the energy and material needs of manufacturing, use, recycling, and disposal of products and services. The Natural Capitalism framework claims that with current, proven technology, we can increase the efficiency of products, services, and other solutions by 30–50 percent. That figure alone would account for a tremendous improvement in our quality of life, as well as our impact on the environment.

While this amount of efficiency would be costly to install, it would also save significant costs in fuel and other resources while providing a major source of opportunity for jobs and businesses. Eco-efficiency is also an important and necessary step on the path toward more sustainable solutions, such as eco-effectiveness (see below).

Here are some examples currently available for solving efficiency issues:

- Putting hybrid engines and drivetrains in transportation of all types

- Replacing both incandescent and fluorescent bulbs with LED lighting

- Using lightweight materials, such as carbon fiber, for weight-critical uses like airplanes

- Encouraging rental, leasing, and borrowing programs to use fewer products more efficiently

- Transforming products into services

- Optimizing the use of material, energy, and water resources

- Reclaiming and reusing waste heat from manufacturing for other purposes

- Recycling all material back into manufacturing

Less Really Is More

Compare these two keyboards (see Figure 5.1). The first is a standard keyboard you would find with most computer systems. It uses a lot of material, most of which is plastic. The second is Apple's new keyboard, which dematerializes the keyboard to a remarkable degree. Not only does it use less material overall, but it also uses less plastic (opting for aluminum instead, which is much more easily recycled). And it still performs all of the same functions.

FIGURE 5.1
Compare these two keyboards.

iPhone

Apple has become a master at dematerialization. In fact, Apple's learned so much about the design, development, and manufacturing processes around this principle that it has become a core differentiator from its competition and has a sustainable competitive advantage. The iPhone is probably the best, though not only, example of this. However, the iPods and most everything Apple produces are also examples of sustainability (see Figure 5.2).

FIGURE 5.2
The iPhone's parts are so carefully designed and tightly packed that there is no extra material to reduce.

To begin, the iPhone has no wasted materials. It isn't designed any bigger than it needs to be or with parts that are used strictly for ornamentation and serving no purpose. It's as small as its components—and the human hand—will allow. In fact, it's nearly a miracle of contemporary engineering. If you were to take one apart, you would be hard-pressed to find any material you could suggest to eliminate—even a bit. This is extraordinary. However, it's neither an accident nor a fluke. All of Apple's products are similarly dematerialized, although the mobile products tend to be to the greatest degree. Apple's desktop computers, like the iMac and Mac Mini, it's accessories, like its Airport

iPhone (continued)

base station, and even its keyboards, use materials more efficiently than all of Apple's competitors' products. To be sure, Apple isn't sacrificing performance with this focus. Its products are some of the most high quality and most technically capable in the industry.

Even Apple's packaging has been dematerialized to the extreme and well beyond most of its competitors (see Figure 5.3).

FIGURE 5.3
Apple dematerializes everything, including packaging.

So how is Apple doing it? The first answer is commitment. Though it's not usually discussed publicly (for fear of castigation or reprisals by environmental groups), dematerialization and material substitution (see Chapter 6) are design directives throughout the company. In addition, this focus extends to the choice of material, not just the amount of material. Apple has moved to aluminum enclosures throughout most of its products. While this material has a higher embodied energy than plastic (meaning, it takes more energy to produce overall giving it a higher impact on the environment), aluminum allows Apple to use less material in more structurally sound designs, which saves weight and bulk and increases the ability and likelihood that the parts will be recycled. (While all materials *can* be recycled, most plastics are in such low demand that they are effectively unrecyclable.)

iPhone (continued)

Apple hasn't stopped there, however. Apple has led the industry in dematerializing its packaging to an extraordinary degree, ending the often shameful waste of packaging for products like software. It isn't perfect (there's really no need to have a box to sell MobileMe Internet services), but it's gone further than any company in the industry in terms of product design and dematerialization (see Figure 5.4).

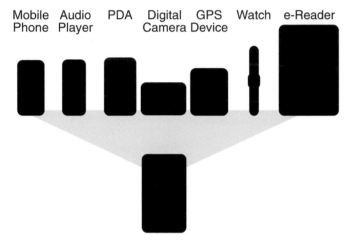

FIGURE 5.4
Some devices are able to dematerialize or reduce the need for all other devices.

Lastly, Apple's iPods and iPhones (as well as some other phones and portable devices) also have the effect of dematerializing other products entirely. When you buy an iPhone, not only are you replacing your phone, but also your portable audio player, PDA (personal digital assistant), and increasingly your watch, digital camera, GPS device, slide projectors (remember those?), and even television. This is dematerialization taken to the extreme, and it creates a much more beneficial impact than merely reducing the materials in a single product.

The concept of dematerialization considers the materials and energy used throughout the product and service life cycle. At each point, designers have an opportunity to make design choices that optimize the use of materials and energy and reduce the impact of their solutions. Some of these choices include the following.

Product Manufacturing

- Create lighter, smaller products, where possible.

- Minimize the number of components, materials, and type of materials.

- Minimize the number and type of fasteners.

- Use materials and processes that require minimal manufacturing energy and materials.

- Use production methods that produce less waste and recycle any waste created.

- Use low-energy production methods and energy from renewable sources.

- Manufacture products closer to customers (requiring less transportation).

- Redesign manufacturing processes to raise quality standards and make quality control easier (so fewer defective products are made or enter the market, only to be returned).

- Develop products to be more useful, usable, and desirable.

Packaging

- Design lighter and smaller packaging.

- Create packaging made from recycled and recyclable materials that are easily separated.

- Create packaging that doubles as promotional material.

- Design reusable packaging (either by customer or by manufacturer when returned).

- Create packaging that doubles as shipping packaging (such as integrated handles or places to put mailing labels so another shipping box isn't required).

- Use materials and processes that require minimal manufacturing energy and materials.

- Create packages of bulk products or concentrated products (like concentrated liquid dishwashing soap or powered soaps).

- Design packaging for refilling rather than replacement.

- Design packaging with labels, graphics, or instructions molded into them that don't require additional labels to be affixed.

Distribution

- Use efficient supply and response systems.

- Plan distribution efficiently based on research of customer use and location.

- Cooperate with other suppliers to optimize shipping and logistics, where possible.

- Plan and specify modes of transportation with lower impacts (such as train and ships over airplanes).

- Create product take-back services to collect, reuse, and recycle products and components when customers are finished with them. Plan for these components to be used in the manufacturing process.

Product Use

- Create products that use energy and materials efficiently (including energy-saving modes).

- Create products that don't require energy for use or allow human power as an option, where possible.

- Specify cleaner and fewer consumables in product-use instructions.

- Allow and specify recycled and reused components and materials for product use.

- Lending and renting programs promote greater use efficiency over a product's life cycle than single ownership.

Services
- Recast product solutions as service solutions.

- Distribute service centers across geographic locations where customers are located, in order to reduce transportation distance for both customers and company service representatives.

- Minimize amount of power used for online and digital service centers.

- Use power from renewable sources for online and digital service centers.

Dematerialization can have one of the greatest impacts in sustainable design; however, it's critical for designers to understand the manufacturing process and the entire supply chain in order to find the opportunities that will have this impact. Manufacturing and distribution aren't processes for others to worry about. Rather, they are critical parts of the solution that need to be considered and potentially redesigned because they represent approximately one-third of the total environmental impact for most products. Designers are seldom taught to work closely with their counterparts in operations (including manufacturing, shipping, inventory, etc.), but these areas are all your potential allies and often hold the information you need in order to design sustainably.

Vampire Power

California has been a leader in rethinking energy use and creating new energy standards for over 40 years. *Vampire power*, the electricity consumed by power converters and appliances left plugged into wall outlets when not being used, is one of the state's latest concerns. Most people don't realize that their mobile phone and other chargers draw power even when the devices they recharge aren't attached. Also, many appliances, such as microwave ovens, televisions, and stereos siphon off considerable power when they're not being used (either *off* or in *standby* mode). This power is consumed without any valuable use. Most of us don't even know it's being wasted, yet it accounts for approximately 10 percent of U.S. household electricity consumption (equal to nearly 114 megawatt hours or over $6 billion worth of electricity in 2004)![1]

Most people don't realize that their mobile phone and other chargers draw power even when the devices they recharge aren't attached.

Much of this energy is redundant, even when it is productive. For example, consider how many appliances in the typical kitchen have a digital clock. It may not seem that these clocks could consume much power (and, individually, they don't), but multiplying the energy consumed by vampire power by all of the appliances in the U.S. accounts for the total output of up to 18 power plants. The typical microwave oven, for example, consumes more power during the day to power the clock than it does when you use it (assuming you use your microwave for under 10 minutes each day).[2]

California's new standards set limits on the amount of electricity that power adapters and chargers draw when "off," and these new standards are over seven times as efficient as older models. These standards are already in effect, even though many manufacturers have been designing them for several years. Similarly, other states and countries are already adopting similar regulations. Designers aware of the issues don't have to wait for regulations to change and can start targeting these inefficiencies immediately.

---

[1] http://en.wikipedia.org/wiki/Standby_power

[2] http://michaelbluejay.com/electricity/vampire.html

CHAPTER 6

Substitution

Designers can have a tremendous impact on the type and quality of materials used in products since they regularly specify materials in order to realize their designs. Often, more sustainable materials can be substituted for less sustainable or even toxic ones without much disruption in the manufacturing process. Sometimes, these materials can also lead to energy, resource, and time savings, as well as reduced liability and risk. However, not all materials can be replaced, particularly in electronic components. Also, different materials will have differing performance and may introduce new issues that affect quality, reliability, and manufacturing. So designers need to work closely with engineers and others in the manufacturing process to be sure that their choices are making improvements everywhere and not merely passing problems from one area to another.

Material Substitution

Material substitution can take many forms and be accomplished in a variety of ways, including the following:

- Substitute reused and recycled materials instead of virgin materials.

- Specify materials that are inexpensive to reuse, recycle, or dispose of and for which there are already systems or networks to do so.

- Use degradable and biodegradable materials where possible, but be wary that these may not degrade as fast or as intended.

- Substitute more sustainable raw materials from more local sources.

- Substitute energy used in manufacturing, distribution, and disposal processes from more renewable sources.

- Substitute water used from more sustainable sources or materials that use less water in their preparation.

- Investigate new and innovative materials and manufacturing processes that may be more efficient and sustainable.[1]

- Avoid laminate films and materials that cannot be separated easily.

---

[1] A great source for understanding the variety of manufacturing processes is the book, *Making It*, Chris Lefteri, Lawrence King Pubs.

- Avoid materials that create monstrous hybrids, combining materials from biological and non-biological sources that prevent recycling.

- Substitute locally available materials and supplies for refilling and reuse to reduce transportation impact.

All materials have different impacts, and it's up to designers to both familiarize themselves with the possibilities and make better choices based on the needs of their solutions, as shown in Table 6.1.

TABLE 6.1

Full Cost Comparison of Many Common Materials	
Packaging Material	Full Cost (per ton)
PLASTICS:	
HDPE	$537
LDPE	$580
PET	$1,108
PP (polypropylene)	$602
PS (polystyrene)	$620
PVC	$5,288
PAPER:	
Bleached kraft paperboard	$443
Unbleached coated folding paperboard	$382
Linerboard	$394
Corrugated medium	$204
Unbleached kraft paper	$390
Folding boxboard from waste paper	$247
Linerboard from waste paper	$256
Corrugated medium from waste paper	$303
GLASS:	
Virgin glass	$157
Recycled glass	$127
ALUMINUM:	
Virgin	$1,963
Recycled	$342
STEEL:	
Virgin	$366
Recycled	$358

Tellus Institute 1992

The Mirra Chair

The Mirra chair (see Figure 6.1), by Herman Miller, is an example of rethinking a product's design around environmental criteria. The inspiration for the Mirra was the success of the Aeron, the (now) ubiquitous chair from the Web boom era. While the Aeron was popular for its ergonomic comfort, its low use of material (compared to traditional, padded executive chairs), and its many adjustments, it was very expensive to build, had many parts, and created a great deal of waste materials in manufacturing. The Mirra is an attempt to keep the popular features of the Aeron while making it a more sustainable alternative. Not only is it 15 percent less costly to build, but its sales quickly surpassed that of the Aeron.

FIGURE 6.1
The Mirra chair is, in many respects, the more sustainable sequel to the iconic and popular Aeron chair.

The designers of the Mirra, Studio 7.5 in Berlin and Herman Miller in Michigan, selected materials with an eye to the manufacturing process and designed the construction to use as few parts as possible. The result is a chair that is easily

The Mirra Chair (continued)

disassembled, uses parts made from 42 percent recycled materials, is 96 percent recyclable, and is even less expensive than the Aeron. Some parts of the chair, such as the back panel, are 100 percent recyclable. Instead of traditional form and fabric, the Mirra uses a natural spring seat of elastomer and a back panel of molded polymer. The designers were also able to substitute nylon for PVC in the cable jacket, and in conjunction with suppliers, specify recycled content in the steel parts. In addition to the material focus, the Mirra is designed to be produced more easily, and at the same time, the simplified production process makes it also easier to both assemble and disassemble.

The Mirra chair is certified at the Gold level of the Cradle to Cradle certification, and Herman Miller is one of the few companies to publish environmental impact reports for all of its products: **www.hermanmiller.com/CDA/SSA/Category/0,1564,a10-c651,00. html**. In addition, all new products developed by Herman Miller go through the Cradle to Cradle process.

Detoxification

There are a variety of materials that have different toxic effects. Some of these are perfectly natural (like arsenic) but nonetheless detrimental. In addition, what may not be toxic for humans may still be toxic for animals (like chocolate for dogs) and plants. Also, any material in sufficient quantity (even water) can be dangerous to any system, especially at the levels created by factories and industry.

Toxins can include heavy metals, formaldehyde, chlorinated cleaners and solvents, particulates, hormones and hormone-mimicking agents, acidic and basic runoff that changes the pH of water and soil, and any number of other categories. Toxins can enter the body through the lungs if airborne, or through the GI track if present in water or food. In addition, many toxins can enter through contact with the skin. Once in the body, the bloodstream can transport toxins throughout the body.

To whatever extent toxic materials can be eliminated or substituted for more sustainable, nontoxic ones, designers will be able to make a difference for the solutions they create. Some materials may not be toxic themselves, but the processes for locating, mining, or processing them may be highly toxic so the sourcing of materials is another issue to consider. The same can be said for the energy used in the processing of materials and components. The more renewable and sustainably-produced the energy is, the lower the contribution a product or service makes to toxicity in the environment and for workers.

PVC as a Toxin

PVC (polyvinyl chloride) is a great example of the toxicity of everyday materials all around us. PVC is practically a miracle material. It is a "thermoplastic polymer" derived from fossil fuel that produces incredibly flexible, durable, and versatile plastic. It can be made into a hard plastic (like vinyl siding or plumbing) or a soft, supple plastic (like plastic watchbands) when phthalates and other substances are added. In addition, it is one of the only plastics that remains durable and flexible when clear. It's also relatively inexpensive, making it a favorite of manufacturers of all kinds of products.

One of the concerns with PVC is that additives (like phthalates) can leach out of the plastic and into food and bodies (through the skin). This is of enough concern that both the U.S. and EU have banned many forms of PVC additives for products like toys (for both children and adults).[2] However, any toxicity of PVC in products pales in comparison to the toxicity created during manufacturing (to both workers and the environment). For over 40 years, scientists have been studying the links between PVC and its related forms, ethyl dichloride (EDC) and vinyl chloride monomer (VCM), in manufacturing and the risks of rare diseases for workers in these industries. In addition, PVC production produces dioxin, another dangerous and persistent chemical, which spreads far distances in the environment. While new processes are being developed

---

[2] www.besafenet.com/pvc

to produce and recycle PVC more effectively, it remains one of the worst materials commonly used in product manufacturing by quite a margin.

The problem for designers is that it is often difficult to substitute other materials for some of the uses that PVC excels in. While it's fairly easy to find substitute materials for PVC's hard plastic forms (though many are more expensive), when designers want soft, clear, or flexible plastic (such as for watchbands and clothing), there are few materials that can take its place—especially at a comparable price. While this may change in the future with new materials and processes, designers can, at least, only specify PVC when absolutely necessary. They can also minimize the amount of PVC used in an application to further reduce the amount of toxins released in the environment or that come into contact with customers.

The problem for designers is that it is often difficult to substitute other materials for some of the uses that PVC excels in.

Some substitutes for PVC include the following.

Plastics

- ABS (acrylonitril-butadiene-styrene)

- EPDM (ethylene propylene diene monomer)

- FPO (flexible polyolefin alloy)

- HDPE (high density polyethylene)

- LLDPE (linear low-density polyethylene)

- NBP (nitrile butadiene polymer)

- PET (polyethylene terephtalate(

- Polycarbonate (PC)

- TPO (thermoplastic polyolefin)

- XLP (thermoset crosslinked polyethylene)

- Ingeo (a new corn-based polymer from NatureWorks) [3]

Non-Plastics (for Decorative or Structural Applications)

- Wood, bamboo, cork, wool, rubber, and other natural fibers

- Cast iron, aluminum, steel, and other metals

More details about the production of PVC are described in the excellent film, *Blue Vinyl.*[4]

Material and energy substitution isn't easy, but it's critical to address. Designers must take the initiative and seek out cooperation with colleagues across the development and manufacturing spectrum in order to make informed decisions about product attributes, as well as giving suggestions to clients on how to evolve. Each project will have specific and different opportunities and challenges, but a short list of tactics you can use applies across most projects:

- Avoid materials that have an adverse effect on human and environmental health whenever possible. This reduces product liability and risk as well as material impact.

- Use the least toxic materials possible and the minimal amounts of any toxic material needed.

- Investigate new manufacturing processes that require fewer toxic materials.

- Investigate processes that remediate toxicity of materials and waste, such as "living machines" or "eco machines."[5] [6]

- Plan to test products, emissions, and waste regularly in the manufacturing process.

- Specifically, avoid PVC whenever possible.

---

3 www.ingeofibers.com

4 www.bluevinyl.org

5 Many plants actually detoxify their environments.

6 www.toddecological.com/ecomachines/

CHAPTER 7

Localization

U ndoubtedly, you've heard a lot about the concept of *local* in recent years. Perhaps, you've read that it's best to patronize shops and companies based in your community or eat food grown as close to you as possible—even food you've grown yourself.

The Power of Local

The power of *local* comes from two ideas. The first is simply that things grown or created near you don't have to travel as much to get to you. This proximity saves energy and emissions since products, food, and services aren't being transported from longer distances. Second, when you support local companies and shops, you're keeping the money in your community (mostly), rather than sending it to other communities (namely, the one where the headquarters for that organization resides). Both of these ideas are smart choices, and the ideas behind them are correct. However, they can have some unintended consequences that aren't always intuitive (we'll address these shortly).

Nonetheless, to whatever extent the products and services you develop can reduce the transportation necessary to distribute, service, and dispose of them, the fewer emissions and other impacts will be generated. Transportation, especially by airplane, is a big contributor to a product or service's environmental impact. Although there are many other ways to measure transportation's impact, one way to measure it is by figuring out what the CO_2 emissions of different forms of travel are, as shown in Table 7.1.

> **... to whatever extent the products and services you develop can reduce the transportation necessary to distribute, service, and dispose of them, the fewer emissions and other impacts will be generated.**

TABLE 7.1

Transportation CO$_2$ Impact (2007), Okala Manual		
Transport Process	**Unit**	**lbs. CO$_2$ Equivalent**
train, long-distance	per person mile	0.013
oceanic freight ship	per ton mile	0.015
train, freight	per ton mile	0.02
train, regional	per person mile	0.024
freighter inland	per ton mile	0.09
automobile, 50 mpg	per ton mile	0.094
truck 40t	per ton mile	0.11
truck 28t	per ton mile	0.12
truck 16t	per ton mile	0.16
container ship	per ton mile	0.17
van, 3.5t	per ton mile	0.19
automobile, 20 mpg	per ton mile	0.24
air, passenger, intercont.	per person mile	0.41
tanker ship	per ton mile	0.61
air, passenger, regional	per person mile	0.78
tram	per mile	0.92
helicopter	per minute	1.6
air, freight, intercont.	per ton mile	1.6
air, freight, regional	per ton mile	2.8

Philip White, May 2007

Obviously, CO$_2$ isn't the only issue, but every mile that can be reduced will lessen emissions (no matter what form they are). But distance is only one of many issues that contribute to the impact of products and services. As we saw in the discussion of Life Cycle Analysis, we need to look at the all of the life cycle impacts in order to make informed decisions. This is where the idea of *local* can get confusing.

Is Buying Local Really the Best?

In 2007, a study by New Zealand's Lincoln University[1] created controversy when it released its conclusion that lamb grown in New Zealand and shipped to England had a lower environmental footprint than lamb raised in England. On the surface, this sounds like a preposterous proposition, and there is still a great deal of disagreement over the study. However, what the study suggested, and acknowledged, was that not every piece of land is as adept at raising food (or manufacturing products) as every other piece. Regional differences can have an impact on that capability, especially in agriculture. The reasoning and calculations used in the study claimed that because England was so developed and its farmland was now so devoid of natural grazing land, in order to maintain lamb, English farms required considerable fertilizers and watering (among other factors) and equipment that used different types of fuels. The summary calculation was that New Zealand lamb accounted for 1,520 pounds of CO_2 per ton while lamb raised in England accounted for over four times that much at 6,280 pounds of CO_2 per ton! The same was also found for some dairy products and fruit.

So, while stores like Tesco in the UK and Whole Foods in the U.S. try to develop labeling that indicates how many miles food has traveled (in order to support the idea of local food), this doesn't tell the entire story. It's not difficult to imagine that Alaska and Norway probably aren't the best places to grow oranges, nor that Sudan has the water to grow rice. The point is that trying to forcibly augment these places in order to reduce food miles will come at a cost. Imagine what it would take, for example, to grow fresh food at science stations in Antarctica. It's not hard to imagine that it may just be easier, with lower impact, to fly the food in as needed (as costly as that is). But, at what point does this trade-off swing in favor of growing or raising food locally?

Remember, this isn't just limited to food. Not every factory in the world is going to be equipped to make an iPhone or a Boeing 787. Some centralization is going to be important if, for no other reason, than to gain economies of scale. Again, it's not clear at what point it makes sense to

---

[1] www.lincoln.ac.nz/story21175.html

decentralize manufacturing, assembly, recycling, and disposal (say, if you have to put anything in a landfill). Only by tracking the full costs and impacts of these activities across the entire life cycle and set of issues can a consumer make an informed decision. It's going to be a long time before we have that data, though.

> It's not difficult to imagine that Alaska and Norway probably aren't the best places to grow oranges, nor that Sudan has the water to grow rice.

But this doesn't mean that designers can't begin making informed decisions. We can't be disheartened by the complexity or lack of answers. We can still arm ourselves with enough understanding to start making better choices. As designers, we also need to stop thinking in absolutes, where we only value the optimal result. Change often comes in small steps, which is okay as long as they're in the right direction. Just because we can't construct the most ideal solution doesn't mean that we shouldn't be making every solution better along the way. That's one of the premises behind the Natural Capitalism framework.

In general, designers can use Table 7.1 to calculate CO_2 impact, and then with data from companies and suppliers, easily and quickly calculate overall impact. This is a good rough guide that, when combined with estimates of other life cycle impacts, can help developers make better decisions. Special circumstances and known regional differences can then be taken into account and weighed along with other factors.

> ... this doesn't mean that designers can't begin making informed decisions. We can't be disheartened by the complexity or lack of answers.

CHAPTER 8

Transmaterialization

*T*ransmaterialization is a strange word, but the process is a new phenomenon not easily recognized by most people. Sometimes called *servicizing*, defined simply, it's the process of turning a product into a service. However, many people still can't understand how this is accomplished without an example. (How *does* a tangible product turn into an intangible service?) So, an example is in order.

The Metamorphosis of Transmaterialization

Consider how people bought music in the past. First, there were records, followed by tapes of different types (reels, eight-tracks, cassettes, and so on), and finally, starting in the 80s, compact discs (CDs). All of these are physical products, even though the music itself wasn't necessarily physical. (It could already be transmitted across radio waves, for example.) Most people associated music with a physical object. Now, however, music is completely digital and even more virtual. The rise in music downloads (both legal and illegal) is displacing the sale of the physical CDs (though some, like records, will probably always be traded by collectors). In this way, the physical product has been displaced by a nonphysical service.

The best example of this is Apple's iTunes music store (see Figure 8.1). Not only does the service enjoy 75 percent of the entire market for music downloads (and sells more music than any other company in the U.S.), but it is also no longer limited to just music. The iTunes music store now sells films, television shows, applications (remember when those, too, were shipped on CD-ROMs?), and even books (as audio books). Where it would have been unheard of to discuss these products as services once upon a time, we now regularly conceive and design new systems that do exactly this.

FIGURE 8.1
The iTunes music store is the most successful platform for distributing digital music, TV, and films.

Of course, simply being a service doesn't mean that there isn't a life cycle impact. In 2003, Digital Europe conducted a life cycle analysis of music CDs bought in stores, bought from Amazon and shipped to customers, and music downloaded from services like iTunes. Their study showed that all of these solutions had an impact of some kind.[1] For example, music services require electricity to run services and lots of equipment to store, process, and transmit their data around the world. It's not insignificant, either. However, even the demands of electronic data are a significant improvement over the demands of producing and distributing light-weight CDs and shipping them around the world.

---

[1]   The environmental and social impacts of digital music: A case study with EMI: www.forumforthefuture.org.uk/node/966.

TABLE 8.1

The Environmental Impact of Different Forms of Music Distribution			
Scenario*	Abiotic Impact (kg)	Biotic Impact (kg)	Water Use (kg)
CDs bought in physical store	1.56	0.09	39.52
CDs bought online, shipped to homes	1.31	0.06	46.73
Music downloaded (user burns CD-R)	0.67	0	23.31
Music downloaded (no CD-R copy)	0.60	0	20.29

* Inclusion of computer materials impact the same in each scenario

The results of this study show a considerable reduction in impact for downloaded music over physical CDs (as we might have guessed). The impact isn't zero, but it's significantly less in all categories. Incidentally, the study also looked at social impacts as well.

Turning a Service into a Product

Digital music isn't the only example where this process has been applied. In fact, rental cars are an old example of turning a product into a service. While rental cars are still physical products, the service of renting or leasing to customers changes both the nature of the transaction and the impact it has on the environment.

When cars are rented from a fleet, they are almost always used more efficiently than those purchased by individuals. For example, rental cars are driven far more than private cars. While, at first, this may not seem advantageous, consider how infrequently many private cars are used. Rental cars offer the potential of using cars more efficiently. For example, renting a large car (such as an SUV or station wagon) for a weekend family trip creates much less impact than driving that same car every day while commuting or shopping (see Figure 8.2). Largely, this is still a matter of social convention since many people buy cars that fit their occasional needs better than their habitual ones.

... rental cars are an old example of turning a product into a service.

FIGURE 8.2
Rental and car-share service cars are used much more efficiently than personal cars.

However, the service of renting cars can also maintain the cars in optimal condition, merely making them available when people need them. The same can be said for recreational vehicles like motor homes, boats, and jet skis, as well as recreational or vacation property. If people are able to rent these easily when they desire, then they don't need to buy and keep their own. For most people, this is not only a more economical alternative, but it's also a more ecological one.

TRANSMATERIALIZATION   145

Similarly, new models are arising as well. Consider the increase in urban car-sharing services, like City Car Share and Zip car. By taking a product (cars, vans, and trucks) and offering it as a service (the right kind of transportation exactly when you need it), these companies are offering people a more efficient and effective variety of vehicles than they could ever achieve individually. One study[2] suggests that car-share services displace between six and twenty private cars for each car in the service, and since customers plan their trips more carefully, customers drive between 50 and 68 percent fewer miles. In effect, these services offer mobility, not vehicles, and that is the conceptual translation that transmaterialization enables. To be sure, however, this isn't always an easy translation for customers to make. Many U.S. citizens still aren't ready to accept mobility as a service, rather than a product. Car sharing isn't for everyone, but more and more customers worldwide are finding that it may be a better use of their money, as well as a better solution for the environment.

> **... these services offer mobility, not vehicles, and that is the conceptual translation that transmaterialization enables.**

Sometimes innovation works; sometimes it doesn't. An example of when it didn't work involves an innovative carpet company, Interface (see Figure 8.3). In the late 90s, Interface revolutionized the environmental impact of one of the worst industries in the world. (You should know that traditional carpet is terrible stuff, from an environmental perspective.) Interface changed everything about the materials and processes they used to make recyclable, less toxic carpet tiles that lasted longer, were better for workers as well as customers, and were taken back by the company when customers were finished with them.

However, when Interface took their ideas one step further and tried to reframe carpet from a product to a service (leasing floor coverings), they

---

[2] www.bbc.co.uk/bloom/actions/carclubs.shtml

weren't as successful. For the most part, business customers just couldn't wrap their heads around the idea that they weren't purchasing carpet, but they were leasing great-looking floors. Even though Interface guaranteed that they would maintain the carpet (including replacing damaged sections) and update, recycle, and dispose of it at their own expense, most people still couldn't make the cognitive leap that they would be paying monthly for something they could otherwise buy outright—even if it represented an overall savings.

Sometimes innovation works; sometimes, it doesn't.

FIGURE 8.3
Interface's innovative solution allows carpet to be more easily installed, replaced, and recycled.

Helping customers think of solutions in new ways is one of the challenges that transmaterialization creates for developers of new services. Services are not foreign to product companies. Each company that sells a product must service it in some way. This service might simply include offering repairs or maintenance. So every product company already thinks about services to some extent, which means that it's not an entirely foreign concept. What transmaterialization offers is a mechanism to rethink the value that those products provide and enhance them by either adding valuable services around the products or by completely transforming the products and their financial models into new solutions.

What Is a Service?

Perhaps it would be helpful to describe some of the components of services. A simple definition of a service is a solution to fulfill needs. The concentration here is on filling needs (and desires) without presupposing a physical product. Even when a product is required, who owns it and how it is used is at the heart of a service solution. To create services effectively, service providers must focus on customer value as an end solution. The aim is to deliver high value (whether in terms of price, performance, emotions, values, or meanings) in customer terms, not merely the terms of the manufacturer. In addition, a priority is often placed on creating and maintaining a deeper customer relationship than is often the case with product manufacturers (who often don't consider anything past the sale of the product).

A service orientation not only benefits environmental and social impacts by optimizing efficiency, but it also often leads to innovative solutions that competitors haven't imagined simply because most organizations operate in a product mentality.

Financial, healthcare, legal, consulting, training, education, and information services are all examples of services that focus on results and value over physical products. In many cases, there are few physical products involved in the service experience. Likewise, restaurants and bars, spas and salons, hotels, rental cars and equipment use materials to create an experience and satisfy needs, but the products are not always the focus of value.

In order to develop services effectively, it's a requirement to rethink the process of deriving solutions. Those familiar with product development will need to augment their understanding with service design fundamentals. In addition, it's critical to prototype services through interaction design in order to validate that important value is actually being provided. Design research techniques that focus on qualitative data instead of merely quantitative data are also important since much of the value provided by services isn't tangible or easily described.

Services may require organizations to engage other organizations as partners in order to provide a complete solution. For example, it's not easy to deploy new information solutions without adequate networks (and creating those may be outside an organization's expertise). Age-old quandaries that plague new technologies often require networks of organizations in order to roll out new services. (Who's going to buy a GPS device, for example, before the GPS network is in place—and vice versa?)

Services can also have an effect on the financial accounting of an organization in unexpected or even counterintuitive ways. When Microsoft announced its original concept of .NET services, it spoke of online subscriptions to its popular software applications (like the Office Suite). They quickly realized, however, that it would destroy their revenues in the near-term since the revenue they make up front, when customers purchase the software outright, would be spread over years instead. This required a change in their strategy and original vision, opening the way for companies like Google, who didn't have such models to cannibalize, to offer them instead.

Services are not foreign to product companies. Each company that sells a product must service it in some way.

Informationalization

P roducing things is both expensive and resource-intensive in terms of material, time, attention, and money. It doesn't matter much what the product is. Whatever can be done to reduce the use for these resources is important, but sometimes it's possible to radically reduce something to almost nothing if we rethink the problem and its context.

For example, while it doesn't cost much to ship a single bottle of Coca-Cola around the world from the original bottling plant in Georgia, multiplied over the hundreds of billions of bottles that Coca-Cola sells each year, that's a lot of financial and environmental impact—even at today's artificially low shipping costs. Long ago, the Coca-Cola company realized that this would be cost-prohibitive for their product and began sending the recipe instead. Using the Coca-Cola recipe and local bottles and ingredients, they can radically reduce the cost of a bottle (or can) most anywhere in the world. In many cases, the ingredients other than water (such as sugar—or now corn syrup, vanilla, etc.) may need to be shipped from their points of origins. However, the amount of material is vastly reduced with this licensing model, and it has allowed Coca-Cola licensees to operate over 600 bottling plants around the world.

> **Whatever can be done to reduce the use for these resources is important, but sometimes it's possible to radically reduce something to almost nothing if we rethink the problem and its context.**

Sending the Recipe

Another example concerns the future of space exploration. Many space enthusiasts maintain enthusiasm for sending humans back to the moon, to Mars, and beyond. However, humans require tremendous resources for supporting life. When humans aren't involved, the materials and energy needed are drastically reduced. Probes, satellites, and rovers have allowed us to explore Mars, other planets, and the edges of the solar system for far less money and materials. Realistically, this will more likely be the future of space exploration, especially because of the distances involved. For sure, it's not the same. It's difficult to get the public excited about a probe when their heads are filled with images of people walking on distant planets. And probes will never be able to do everything that humans can (though they can do many things humans can't). However, the material requirements will reduce human exploration, and this is a model for how other solutions can be made more sustainable.

Already, some products are disappearing into bits from atoms. The digital music example of transmaterialization in Chapter 8 also serves to illustrate the strategy of informationalization. In fact, without the ability to faithfully reproduce music digitally, it wouldn't have been a possibility to turn music products into music services.

Email is another good example of informationalization. While we can't send everything through email that we once sent via physical mail, most of what we communicate doesn't require physical material—especially business correspondence where sentiment isn't as important—opening the opportunity to communicate digitally at a vastly lower environmental impact.

Informationalization is all about sending the message, the recipe, the data, whenever and wherever the physical thing itself can be replicated at the destination. The history of horticulture has been a story of this process. It is one thing to send the fruit or tea or spices from a faraway land, but another to send the plant itself—or, better yet, the seeds—in order to provide a continuous supply of the material. Where it simply wasn't practical in the past to send the plant or fruit (forcing us to send seeds instead), now, it's getting more and more costly to send the product than the recipe. Of course, it's also less sustainable to do so.

Email is another good example of informationalization.

Taking the Recipe Concept Even Further

What other things can we send the recipe for instead of the object?

One solution is in the advances made by rapid prototyping machines. We already have the ability to print books at the source, instead of at a central printing facility (think downloadable PDF and ink jet printer). Why not do the same with furniture, dishes, machine parts, and other products? Rapid prototyping processes (and the machines that employ them) already do this for some applications (such as prototypes). Currently, these machines are limited in what they can produce, and the cost of the finished object is much more expensive than other forms of production. However, as the price falls for both machines and materials, at some point these devices will be affordable for people to have in their homes (or at a local site).

Perhaps the best example of informationalization is the biology of nature, in particular, DNA.

There are a few different rapid prototyping processes, each employing slightly different materials and methods. Some use a powdered plastic material, others a liquid. Some are capable of producing hard plastics, others soft (like rubber). Some can even produce objects in color. There was a time that having a laser printer at home, or a color printer, was too expensive for most people. There may be a time when, like color printers now, rapid prototyping machines will allow us to send just the instructions to produce a design, rather than the finished object itself. This will drastically reduce the transportation impact, and because these processes are usually very efficient, possibly the production impact as well for many product solutions. Designers continually need to stay abreast of new manufacturing processes in order to take advantage of advances that affect how and what they can realize. These technologies offer the potential to build things that cannot be built in other ways while also reducing their environmental and possibly even social impact.

Perhaps the best example of informationalization is the biology of nature, in particular, DNA. Consider how efficient this material and process can be: one cell, carrying a nanoscopic bit of molecules, can be responsible (in the right environment) for creating an entire organism. Even a simple virus, carrying a bit of DNA code, can splice into the replication process to insert itself, be replicated, and slowly change an organism. As well as being an example of how nature has, once again, beat us to an efficient design strategy (see Biomimicry in Chapter 3), this example may provide a new kind of design process and solution.

Since the 1980s, science fiction writers and some designers have been hypothesizing that the design of complex objects, like cars and buildings, can be produced via DNA code. Imagine a small set of DNA code that, when inserted into a vat of biological material, will code-assemble a complete car—structure, electrical system, engine, body, interior, and all—simply by chemical instruction. Or consider nanotechnology and the billions of dollars spent on research into nanoassemblers—submicroscopic machines that can build products one molecule at a time. If these visions come true, the roles of designers and engineers will change radically into something more like chefs developing a recipe. In the process, the materials and energy required to produce such solutions may become efficient to the point where they begin to approach the efficiency of nature's exhibits.

> **Imagine a small set of DNA code that, when inserted into a vat of biological material, will code-assemble a complete car—structure, electrical system, engine, body, interior, and all—simply by chemical instruction.**

Architecture for Humanity

In 1999, Cameron Sinclair and Kate Stohr started an organization, Architecture for Humanity,[1] as a response to the housing needs of refugees in Kosovo. Dissatisfied with the mundane demands of his current architecture job, Cameron reasoned that his skills—and those of other architects—could have more impact helping those in real need. He recognized that while food, water, and medical needs were a primary focus, shelters for refugees were unnecessarily costly, ill-suited to the locations in which they were deployed, and too few to respond quickly for the amount of need. And this was a perpetual problem (see Figure 9.1).

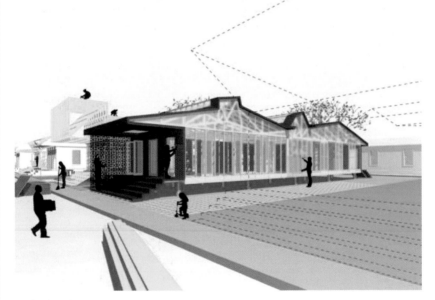

FIGURE 9.1
Architecture for Humanity's Open Architecture Network makes plans for all kinds of temporary and permanent structures available to everyone around the world.

[1] en.wikipedia.org/wiki/Architecture_for_humanity

Architecture for Humanity (continued)

Armed with this understanding and drive, the two quickly organized a small competition to design and engage architects. They rightly assumed that, if they had ideas for better (medium-term) transitional housing solutions, and could engage others in the process, still better solutions could be created. This was the first stage of using the strategy of informationalization to better solve the temporary housing challenge (though the two would probably have never used this term).

The response was overwhelming. Over 220 entries to the competition were submitted, from 30 different countries—all with only a minimum of awareness in the industry. People, of course, weren't sending actual housing prototypes, merely renderings, descriptions, and specifications. Because the field of architecture has developed enough language and conventions for evaluations, Kate and Cameron—and the other judges they had assembled—could make a reasonable critique of the entries without having to build each one. Another key learning fact that emerged immediately was that there wasn't a single, *best* solution. In fact, the solutions were surprisingly ingenious and suited to different uses— even more than Stohr and Cameron had envisioned at the start of the competition. What they had amassed was a catalog of many, outstanding solutions. The next question was "what to do with them?"

AFH quickly learned that the realities of the relief industry, which were scouting suitable locations, getting materials through customs, securing work permits and approvals, etc., were next to impossible. They were not going to be able to select one or even a few of the structural designs to be created in mass quantity, get materials where they were needed, or have housing built. They needed a different solution. So, the two did something different than standard operating procedures in the relief industries. The competitions had been so successful, and had taught them the power of bringing multiple minds to bear to a challenge, so in 2007, they created the *Open Architecture Network*. Employing "Creative Commons" licensing, this allowed any designer or architect to upload any solution to the network, and it could be downloaded by anyone who needed it wherever they were. In effect, they created a recipe catalog for relief housing accessible from anywhere with an Internet connection. The catalog is open so the recipes are ever-

Architecture for Humanity (continued)

expanding, in order to meet the ever-expanding needs of refugee, relief, and temporary housing needs.

The end-solution requires some management and tools for sifting through the designs to find the best and most appropriate way to make the system more usable, but it is an extraordinary solution and one that thrives on the idea of informationalization.

Reuse

We've all seen the surveys that rank lawyers and used car salesmen near the bottom of trusted job titles. In their defense, however, used car salesmen are providing a service that serves sustainability. It is almost always better to reuse products and components than to dispose of (or recycle) them in order to make room for new ones. This principle, of course, relies on the fact that the older products were well designed, met our needs, and were socially and environmentally acceptable.

Consider buildings, for example. Our desire to demolish older buildings in order to build bright, shiny new ones (that rate highly in LEED[1] building standards) often creates an amount of wasted material and energy that the new buildings could never compensate for. Destroying all of the structure and material in a building, plus hauling away the remains, plus transporting the new materials to the site, plus the new construction takes a lot of material and energy. It's often a more sustainable solution, providing that the building is sound, to retrofit existing buildings with new materials (only where necessary) and efficiency technologies. Likewise, continuing to drive an older Honda is often better than scrapping it and driving a new hybrid.

Designers and engineers often dislike these solutions because they feel deprived of the ability to create something new. Plus, they fear that there won't be a viable market for entirely new solutions if people reuse every piece of older ones. (Cuba and its reuse of classic cars is a notable exception.) There's little chance that such a shift in society will take place quickly, so we'll all still have jobs for the foreseeable future. However, if we care about sustainability, we have two new opportunities and responsibilities. The first opportunity is to retrofit existing structures and solutions whenever appropriate, which is in itself a design and engineering

---

[1] LEED: www.usgbc.org/leed/

challenge. The second one is to rethink the reusability of everything we develop from now on.

Careful understanding of our customers' contexts, needs, and desires, as well as an understanding of the ecosystems of materials, energy, and life cycles of the things they use, can lead us to better solutions for the new things we do bring into this world and what will happen to those things when they are no longer needed. Not only can reusability be designed into some products and services, but also the ease at which things can be reused is directly a function of how we design, engineer, and produce things.

We can't anticipate all reuses of products, of course, nor should we try. For example, Artechnica takes wine bottles and transforms them into beautiful vases simply by cutting their tops off at an unusual angle and finishing the edges (see Figure 10.1). The designers of the original bottle could hardly anticipate this use of their product, yet the end result is still something beautiful—and more valuable—than the original empty bottle.

However, we can keep in mind the reuse of the things we design by understanding how they're made and used, and how they can be valued when our customers are finished with them. One strategy is to extend their use in the first place, making them last longer. Another is to design their operation and components to be easily exchanged so that the majority of the components—and environmental impact—stay in use. By consciously making components easy and cost-effective to replace, many products can live on much longer than we would expect.

FIGURE 10.1
Artechnica's tranSglass® vases reuse common wine bottles to uncommon effect.
www.artechnicainc.com

CHAPTER 10

Design for Durability

Anything that developers can do to extend the life of a product or service reduces its resource and environmental impact simply because it doesn't require a replacement as soon. *Quality* is one way to accomplish this. (However, this can be problematic for products that respond mostly to short-term trends or where customers are unconvinced that the extra expense is worth the higher price). Another approach is to make the product easily repairable so that most of the parts (and those with the highest impact) can continue to be used. This is difficult for products whose technology is evolving rapidly, but often even these products can be designed with modular assembles that allow some parts to be upgraded while others (such as cases) continue to be used.

Anything that developers can do to extend the life of a product or service reduces its resource and environmental impact simply because it doesn't require a replacement as soon.

Planned Obsolescence—The Downfall of Durability

The term *planned obsolescence* is the second most dangerous concept ever invented by marketers. (The first is *retail therapy*, discussed in the conclusion.) It encourages us to give up or throw away perfectly good things in favor of others simply because we're led to believe that they're no longer useful or fashionable. The phrase was coined in the late 1920s, but it was popularized by an industrial designer, Brooks Stevens, in the 1950s and found particular popularity in the automobile industry in convincing Americans to purchase new cars more often. It became a major driving force for American cars and car buyers, who became so fixated on trends and styling in cars that they upgraded frequently. Instead of inventing new features that would improve people's lives (how much longer did we have to wait for intermittent wipers, cup holders, and release latches inside trunks?), car designers were told to focus on dramatic, often outlandish, and sometimes even dangerous styling elements.

Now, as I said in the "Introduction," styling isn't bad. Styling is often the expression of our own individuality or our identities. The problem here was that car companies made unsuspecting people think they were all but un-American if they drove a car that was more than four or five years old. Styling during this time was more about ornamentation than fashion or design. It fed on insecurities—and even created them —rather than satisfying people emotionally or meaningfully.

> The term *planned obsolescence* is the second most dangerous concept ever invented by marketers.

I'm not suggesting that people shouldn't be able to buy something fresher, newer, or more appropriate. People's tastes evolve. Wine collectors regularly cull perfectly great wines from their collection that no longer hold their interest, appreciation, or imagination. But they don't throw away these valuable wines, but rather sell them to other wine enthusiasts who will appreciate them. They wouldn't think of simply opening the bottles and dumping them down the drain.

Fads Versus Trends

For designers of products that rely on fads, design for durability is the antithesis of their strategy. It's probably not realistic that fads will go away—or that customers will stop responding to them—so it's important to know the difference between a *fad* and a *trend* and how to develop accordingly. If faddish products are going to be produced at all, then by all means, they should be made to be recycled and disposed of easily—more easily, in fact, than all other products. If we know that a particular "faddish" product isn't going to be around for long, then let's not design it from materials that will prevent recycling. That's called *value engineering*, and it's a perfectly acceptable endeavor, as long as it's not the driver of the solution. When decisions about materials, processes, and configuration are driven by a desire to decrease product life, then something is wrong, and we're planning for obsolescence.

Forms of Obsolescence

Obsolescence takes different forms: technological, aesthetic, functional, and cultural.

Any of these may be planned, but several of them (particularly technological and aesthetic) are often natural evolutions that organizations can't ignore. Technological development is often necessary to improve solutions, whether in production, use, disposal, and so on. Imagine, for example, if a technology were finally perfected that reduced power consumption by half while improving performance threefold. Should companies *not* release products with these new capabilities simply because they require some translation, learning, or adaptation from current devices?

While we may bemoan the fact that the computer, television, phone, or iPod we bought two years ago now seems obsolete, it's not likely the case. Most of this feeling is only based on recently introduced versions with new features or performance. These products still work exactly the same as they did when we first bought them and almost always still satisfy our functional needs. If the new features were truly needed, then we might actually be experiencing functional obsolescence. If they weren't (which is usually the case), then it is cultural obsolescence (driven by our emotional and not functional requirements).

While we may bemoan the fact that the computer, television, phone, or iPod we bought two years ago now seems obsolete, it's not likely the case.

Some things we buy—electronics and digital devices, in particular—do become less functionally appropriate quickly. That's a reality of fast-developing technology. It's not that a laptop or mobile phone doesn't do the job anymore, but that the systems it relies on have marched forward with progress, often stranding a perfectly good device that otherwise would have been useful. Typewriters, for example, work just as well as they always have. You can take an old, cared-for typewriter, put in a fresh ribbon (if you can find one) and a piece of paper, and type out a letter as reliably as you could for the past century. Not so with a desktop computer from even 10 years ago that runs with an operating system no longer supported and no physical connectors to a device or network usable today. Most schools won't even accept donated computers more than three to five years old, now. While I understand all of the reasons for this, it doesn't *have* to be this way.

We can fault companies for intentionally contributing to cultural obsolescence (and ourselves for falling for it), but we can't fault them much for technological obsolescence, since this is an important part of innovation and continuous improvement—not that this makes us feel any better about the constant march forward in technology.

We Westerners live in a supremely wasteful society, and we're promoting our way of life, already unsustainable as it is, to others around the world. Most of what we buy and use, we don't use up—at least not the expensive, complicated things with the most impact. That's fine if we design a system to deal with these partly used, often serviceable products. Some things get better with time, like wine (at least to a point). Some take on comfortable modifications that make them more valuable to us or exhibit their enduring, reliable usefulness. Others can find other homes. Still others have perfectly good components that don't need to be scrapped along with the ones that have worn out or stopped working.

If we can find ways of creating products and services that meet needs over longer periods of time, we can save materials and energy. For example, doubling a product's useful life saves the materials and energy required to produce an entirely new product to replace it for that second half of its life. That's a significant savings. This isn't possible for all products, though. The smallest, portable goods are often the ones most tightly engineered with decisions that decrease lifespan in favor of increased efficiency or more portability. As we dematerialize products, this becomes an increasing challenge, but we have a long way to go before dematerializing small products conflicts with designing them to be repairable or longer lasting.

How to Design Products for Sustainability

Sustainable designers look for ways to design appropriate components that can be exchanged so that the product as a whole can endure longer. Perhaps a mobile phone's case gets marred over time, having been dropped too many times or from swimming in a purse or pocket with keys, coins, and other hard-edged objects. There are several potential solutions here that could increase the life of the phone:

- Make the case out of more durable, scratch-resistant, or colorfast materials.

- Make the case out of materials that age gracefully or create a desired patina (like wood or denim).

- Make the case easily exchanged so that the rest of the phone lasts longer. This has the added benefit of enabling customization and protects the components from the whims of trends.

- Standardize components so they can be used in many different models, increasing the lifespan of the subsystem even when the software, casing, or other components change over time.

- Design devices to be expandable, even just a bit, allowing other devices to be added or attached as they become available or change.

- Transmaterialize the functions, where possible, so that the physical device can outlast changes in features and interface without requiring a redesign.

Strangely, solutions like snap-on cases of different colors and patterns have been around for years now, but still many phone models are designed without them. This is a shame.

Design for Durability

In the world of interfaces and interaction design, cognitive models (sometimes metaphors) change rapidly as new ideas arise. This rapid change often creates fatigue and even cognitive dissonance in users who have to relearn the use of a device, a Web site, an application, or even a system. Some strategies for reducing this problem and lengthening the lifespan of these devices and services include the following:

- Design interfaces and systems with future needs in mind wherever possible. Consider total customer needs and design models that take these into account, even if not all features will be available in the beginning.

- Design an effective, usable system to begin with. Solutions that aren't usable (which may be discovered only after the product is purchased) often are discarded, entering the waste stream prematurely without providing any value for the energy, time, money, and materials used to create, ship, and sell the device (see Chapter 4, "Design for Use").

- Choose metaphors that won't get old quickly (or don't use one at all).

Design interfaces and systems with future needs in mind wherever possible.

Choose Components Carefully

Along with increasing the durability of products by choosing higher-quality materials, fasteners, and manufacturing processes that last longer, developers can also identify and eliminate defects and weaknesses that would otherwise prevent a product from working for a long time. Extensive testing can identify some of these, but others can only be witnessed while in use by customers (who often store, use, and maintain products much differently than their manufacturers intended). Close relationships with customers can help track product defects and failures that might appear on other models. (Note that the airline industry does this meticulously.) Also, extended testing of products can generate important data on product durability, both inside and outside the company. (For example, some automobile companies give magazine reviewers their car models to drive for several years in order to review them over the car's lifetime.)

Where possible, when broken components of a more complex product can be exchanged for a new replacement, the result is a more sustainable solution that minimizes material and energy impact. Imagine having to junk a car when the battery died or throw away a printer when the ink ran out? As silly as these examples sound, there are plenty of products that aren't designed for replacement of components, requiring perfectly fine undamaged components to be recycled or disposed of unnecessarily. Apple's iPods and iPhone are examples of this and mar otherwise admirable and advanced sustainable designs. Most televisions and smaller kitchen appliances (everything but dishwashers, stoves, and refrigerators) suffer the same fate. When one component fails, such as the pump on a coffee maker, it's usually impossible to get a replacement part, or the labor and cost involved in installing it far exceeds the value of the appliance itself.

Pay Attention to Serviceability and Maintenance

Component replacement is part of a product's serviceability or *maintenance*, including easy care by the user or owner and easy and effective care by service people. Serviceability requires some systems thinking and a great deal of understanding about how users and other customers work with the product (and the system in which they work). It's critical for maintenance to be easy and effective, including the opening and closing of the product in order to replace parts without breaking either the product or the warranty. Designers and engineers must work together across more than just scenarios of manufacturing and use, specifically developing for scenarios of maintenance and misuse. I have yet to see the results of user or design research that include profiles of undesirable users or scenarios of misuse—or those where users require help. Yet, these are often the most critical scenarios of all, since they are the places where catastrophic problems or decisions may arise.

Designers and engineers must work together across more than just scenarios of manufacturing and use, specifically developing for scenarios of maintenance and misuse.

Design for Upgradability and Customization

If components are designed to be easily replaced, they will be more easily *upgraded* as well. You'll find this fact to be especially true for components that wear, as opposed to those that are outdated due to technology. Upgradability also enables *customization* and *personalization* to be more effective, and these are two attributes that also contribute to overall product life. If we can configure solutions to better meet our needs, then we're much less likely to simply drop them for something that isn't as personalized. Plus, the personalization process can increase the usefulness of many products, not just the desirability factor alone.

Upgrades aren't easy with lots of technological products, even though they're even more appropriate since the evolution of technological change is so rapid. While larger, personal computers (such as "towers") and servers are almost always designed for interchangeable parts and upgradability of major components, smaller computers (like laptops) are often engineered to such tolerances that component upgrade is sometimes at odds with other requirements. These products are so tightly engineered that changing a processor, for example, might not have much effect without also changing the motherboard, other processors, and other components in the system. What might sound like an easy and obvious upgrade ("just switch the processor for a faster one") often isn't easy at all when the whole system is considered. Still there are opportunities to increase the upgradability of laptops, iPods, and mobile phones. Most, in fact, already allow users to exchange SIM cards, batteries, memory, and other components. But designers and engineers need to prioritize the most critical needs of users and find ways of increasing reliability and longevity of devices around these needs.

> I have yet to see the results of user or design research that includes profiles of undesirable users or scenarios of misuse—or those where users require help.

In some cases, a new business model might lead to longer-lived products. For example, a rental or leasing system for products might maintain them better regularly, leading to longer life. Services centers could be stocked better to service community or regional products where individuals might not be able to afford the tools or parts on their own. The better products are maintained, and the more that this happens locally, the larger the sustainability reward (and lower the environmental impact).

Project Better Place: Making Electric Cars Practical

Shai Agassi, founder and CEO of Project Better Place,[2] is building a solution for Israel (and now Denmark) to completely change over to electric cars by 2020. This is an extraordinary goal, even though these are both small countries. Part of the strategy to make electric cars more effective for people is to give them the option of either recharging the batteries or replacing them at a service station within five minutes. To make this possible, customers will buy the car, but lease the batteries. In essence, the lease gives customers the right to have a fully-charged battery in their car at all times. For those driving long distances, this makes electrical cars practical. Of course, it also requires an extensive network of service stations able to swap the battery, which is a lot easier in smaller countries

FIGURE 10.2
The Better Place Rouge, a car produced by Nissan, should come soon to Israel, Denmark, Hawaii, and the San Francisco Bay Area.

2 www.projectbetterplace.com

Project Better Place: Making Electric Cars Practical (continued)

like Israel and Denmark but still practical for even larger countries, like the U.S. (see Figure 10.2). In this case, the model of leasing a component allows the electric car to compete against gasoline cars favorably. Otherwise, they are at a disadvantage.

The Value of Redundant Components

Some products and services are more durable if they include redundant components. Redundancy is especially important for critical systems. Often, these solutions can't simply be taken "offline" or turned off. Failed components may need to be changed while the solution is active. Designing for this kind of durability can also make it easier to upgrade critical systems while they're active as well. It may also help them to be more flexible in terms of use and in responding to technological change.

Create Classic Styles

Designers can frequently increase the durability of the solutions they create by designing in classic styles, rather than trendy ones. These design styles include decisions of form, color, typeface, texture, material, and so on. Typically, design that adheres to classic tastes will meet more people's needs for a longer period of time, although it is at odds with the trendy and faddy nature of much of the design industry and media. Consider the first iPods (now called the *iPod Classic*). When these first appeared, the media called them a beautiful design, but most designers looked at them and asked "where's the design?" To graphic and industrial designers alike, they could hardly be less designed: a white rectangle with rounded corners, no ornamentation, no color, white circular controls, and a rectangular screen. Compare these to the many MP3 players of the time from Rio, Samsung, Sony, and so on that came in all manner of shapes, colors, and materials (see Figure 10.3). Now look at the marketplace: few of these MP3 players remain viable (and customers aren't clamoring for other shapes and styles, though color choice is still desired), and most competitors have moved to more classic solutions.

FIGURE 10.3

MP3 players take a myriad of different forms but many seem faddish and destined for fast retirement.

Remanufacture Products

Products designed to be remanufactured also help increase product life. Modular structures can help make repair and replacement easy and effective. Products that fit a range of uses (like the iconic Swiss Army Knife) can sometimes outlast those with specialized uses (or misuses).

More than anything else, however, products that are meaningful (that resonate with our values, emotions, and meanings) are often the most satisfying and durable of all. To whatever extent you can develop products and services that connect deeply with customers, the likelihood that your customers will keep these products longer increases dramatically.

The Dyson Vacuum Cleaner

When James Dyson became so frustrated with the conventional vacuum he owned and decided to design something better, it was hardly his first invention. In the late 1970s, the industrial designer decided that a different process altogether could both cut down the number of moving parts, provide continuous and better suction, and not lose that suction at all as it picked up dirt like traditional vacuums did. His dual cyclone vacuum used the principle of cyclonic separation to suck up dirt and drop it into a clear bin (so you could see it) without clogging the motor or requiring a bag (see Figure 10.4).

By eliminating the bag, the vacuum cleaner eliminates a lot of material and impact in the use phase. In addition, the vacuum can save energy since its suction isn't diminished during use. Lastly, the reason why this is an example of durability is that the vacuum itself performs better for longer since it doesn't rely on costly bag replacements that might, someday, cease to be produced.

FIGURE 10.4
The Dyson vacuum cleaners use strategic design to create more durable and effective solutions.

Design for Reuse

It's not enough anymore to simply design better, more durable products. In order to be truly sustainable, solutions need to both last longer *and* have a life after their normal use period. Not every product can be reused as something other than its original design, but many could, if their designers only considered this possibility in the development process.

Like extending product life, designing products to have uses past their normal, intended use keeps their materials outside of the waste stream. Reusing a product's materials (such as components or waste) is one approach. For example, could a used tire be put to use in better ways than burning? However, reusing a product itself, when its normal usefulness is past, is even better. Not all secondary uses can be designed as part of the original solution, but innovative approaches can make it easier for one product to be used as something else later.

It's not enough anymore to simply design better, more durable products. In order to be truly sustainable, solutions need to both last longer *and* have a life after their normal use period.

Two Types of Reuse

There are two types of reuse: *deliberate* and *unintended reuse*. The first occurs when products and services are reused in a way that is expected and intended. *Deliberate reuse* can span many types, including objects that are reused in similar ways as the original intended usage and those that are planned for completely different uses. *Unintended reuse*, such as the Artechnica bottles shown in the preceding pages, can't really be anticipated by developers. It's fantastic when it occurs, but it's not something most designers should spend time pondering about (aside from the strategies of dematerialization, making their products durable and easy to disassemble, and so on). *Make* and *Ready Made* magazines[1] are great sources of ideas for unintended reuse.

---

[1]   Make Magazine: **www.makemag.com**
Ready Made Magazine: **www.readymade.com**

Think Long-Term Use

Making a solution more durable is, in effect, a type of reuse, but this chapter is about extending past that type. Instead, designers should consider if and how they can transform a solution from once or short-term use to long-term use. Anything sold as "disposable" might better serve customers and sustainability alike if the solution incorporated a reusable context.

> There are two types of reuse: deliberate and unintended reuse.

For example, a disposable one-use bag might be convenient, but a reusable bag (made from more durable materials) would be more sustainable over time. There may be a place for disposable diapers because of their convenience, but a scenario of reusable diapers that are laundered is likely more sustainable. To be sure, not everything can or should be reused. Some medical equipment is designed specifically to be used only once and not to be cleaned. In fact, this is now the standard for syringes, needles, and many other medical devices. However, while dressings and bandages shouldn't be reused, for years many medical products, like IV bags and needles, were cleaned, sterilized, and reused without any problems.

> Anything sold as "disposable" might better serve customers and sustainability alike if the solution incorporated a reusable context.

Maille Condiments

For decades, Maille, a French manufacturer of condiments, has been selling mustard, mayonnaise, and gherkins in glasses designed to be collected and reused as drinking glasses in the home (see Figure 11.1). Although known for its tasty, premium mustard, the company is as well known in France for its approach to reusing glass jars. The seal and cap are designed for easy removal after use, and the edges are already finished for use as drinking glasses. The company often releases special commemorative series, and the latest glasses are a surprisingly premium design.

FIGURE 11.1
Maille has been selling its condiments in reusable jars for decades.
www.maille.com

In our current, throw-away society, there are precious few examples of purposefully designed reuse. In the 1940s and 1950, when conservation was *de rigueur*, this was more common. However, there's no reason why more products, especially consumables like food, water, paper products, soaps, cleaners, and so on can't be designed for reuse in other ways. Packaging is an especially good opportunity for product reuse.

One way to help products be more reusable is to make them *easy to clean.* Another way is to design them to *separate easily from packaging* or to use components that aren't designed for reuse (for example, the lid of the Maille jars). Modularity and interchangeable parts improve product reusability, as do simplification and standardization of parts and components.

Energy, too, can be reused in innovative ways. For example, several years ago, lampshades for children's rooms started appearing that used the heat coming from an incandescent bulb (for people still using those) to make the lampshade turn around, simply from the convection of hot air rising through the vanes in the lampshade. The lampshade, then, became a spinning stencil for projecting light figures in a room, like a mirror ball with shapes.

Even though it's nearly impossible for designers to plan for unintended reuse, designers can often develop solutions that recycle items in completely different ways. Technically, this is still recycling, so look at the example in Chapter 13 for William Good.

Rapioli™

Moving parts around the world is an important business function, but the packaging used to ship them is often discarded and never reused. To make a more sustainable solution, Ken Eskenazi created a brilliantly designed system, called *Rapioli*, with modular components that allow packaging to be reused over 100 times and easily stacked and stored (see Figure 11.2). Intended for the business-to-business market, the packaging is made from only two reversible parts (the outer shell, and an inner cushion) that consist entirely of recycled PET plastic.

In addition to reducing waste, reusing excess PET, and reducing the need for paper, Rapioli also saves companies money on each shipment, compared to standard cardboard boxes. Material labels are molded into the packaging, along with flanges that snap the two shells together and nubs to aid stacking. There's ample room for packing labels as well.

In all, the solution's ingenious combination of material selection, industrial design, and market-appropriate economics makes it both more effective and more sustainable.

FIGURE 11.2
The Rapioli reusable packaging system.
www.innovation2industry.com

Recycle

We've all heard about the need to recycle, but few of us do it to the extent that we could, and this causes a great deal more environmental impact than necessary. For example, in 2005 (the latest year for the EPA to report), U.S. citizens recycled about a third of what went into the municipal waste stream. That's a steady improvement, but not a figure to be proud of, nor a sustainable strategy. The items with the best recycling rates included containers and packaging (roughly, 40 percent). In addition, approximately 60 percent of yard waste was recycled overall, although far less food waste. However, to temper all of these gains, the overall amount of trash is increasing rapidly. From 1980 to 2005, U.S. citizens generated more than double their amount of trash[1] (see Figure 12.1)

Some communities already recycle at much higher rates. (San Francisco hit the new record in April 2008 of 70 percent and is focused on improving even more.[2]) These efforts are important because they provide models for other communities to follow, at the governmental, citizen, NGO, and business levels. For example, San Francisco's system involves tight coordination between the city and municipal waste collection, sorting, and recycling facilities. In addition, education and convenience for customers are critical to these high rates. There's also a cultural force at work within the community that creates the desire and understanding of recycling.

---

[1]   www.ens-newswire.com/ens/apr2008/2008-04-24-092.asp

[2]   www.epa.gov/epaoswer/non-hw/muncpl/pubs/mswchar05.pdf

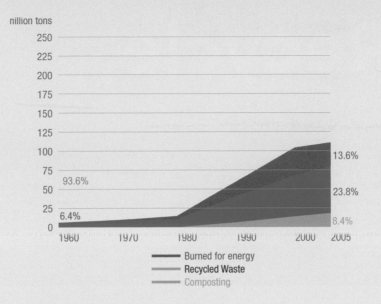

FIGURE 12.1

While the total waste generated by communities is rising, so is the percentage of waste that is recycled.

CHAPTER 12

Design for Disassembly

Recycling is an important tenant of sustainability, but in order to be effective, products need to be easily disassembled into component parts and separated by material. If this is difficult, these products simply end up in the landfill instead.

The worst parts, in terms of recycling, are those made from two different materials bonded together, because they can't be easily separated. The Cradle to Cradle framework designates these as "monstrous hybrids." A good example of this type of hybrid would be milk and juice cartons that come with circular pour spouts and caps built into the side. The plastic cap and spout can't be recycled with the waxed cardboard, and yet there are no easy ways for recyclers to separate these quickly. While this design is particularly convenient for some users, it makes recycling nearly impossible (a good example of opposing goals). The only way to recycle these is for users to cut the plastic spout from the rest of the container before placing them both in a recycling bin.

Likewise, most modern clothing presents a particular challenge since so much of it is blended from natural and artificial fibers. That shirt with 80 percent cotton and 20 percent rayon can't be recycled in either the compost or recycling bins. Currently, it can only go into the trash bin (and, thus, the landfill) since we have no economical way of separating the two materials.

The next most difficult materials to recycle are products that can't be taken apart easily (therefore, they go into the trash instead of being recycled). Any product with parts that are bonded together or sealed so that they can't be disassembled at all are not going to be recycled. Likewise, products assembled with complex assemblies or requiring custom or multiple tools aren't likely to be recycled either. If it takes too long for workers to disassemble, they just won't do it. In addition, these same products likely won't get repaired often (or ever) for the same reason, which makes them difficult to endure or reuse. Design for disassembly is often the same as design for assembly. Improvements that designers and engineers make toward the end goal of ease of disassembling a product often take into account—and improve—how that product is assembled in the first place.

Disassembly, Step by Step

It's not too difficult to design more easily disassembled products when it's part of the initial phase of the design specification and goals. However, once engineering, design, and production are already decided, it's nearly impossible to redesign for easy disassembly.

To whatever extent possible, designers and developers can increase the likelihood of their products being recycled by using the following techniques.

Pure-material parts: These are parts made from only one material that doesn't need to be separated. For most products, it's unlikely that the whole product can be made from the same material, but if a product's parts are at least uni-material, then each can be recycled easily.

Fewer parts: Where possible and applicable, reducing the number of parts can reduce the time and cost of disassembly and also affect the overall environmental impact by potentially reducing the amount of materials used.

> Have you ever tried to assemble a piece of furniture or tried to repair an electronic product only to find that there were many different fasteners used throughout the product?

Batteries and other electronics that are easy to remove: These components are often the most hazardous in terms of toxic chemicals, and they should be separated from the rest of the waste stream as early as possible. Ideally, they should be recycled separately, which isn't possible if users can't pull them out of the product easily.

Standardized fasteners: Have you ever tried to assemble a piece of furniture or tried to repair an electronic product only to find that there were many different fasteners used throughout the product? It sometimes feels like each bolt or screw has a unique length, size, and head configuration, needing a unique tool to deal with each. While this is often a deterrent on the part of the manufacturer to prevent users from repairing their own products, it also increases the complexity (and likelihood to mistakes) in terms of assembly and repair for the manufacturer, too. This is just one of the many other benefits that standardization can create.

Accessible fasteners: Anyone who has ever tried to change the oil filter on a typical small car, or a fuse in just about any car, knows the pain involved when parts aren't easily accessible. There's no reason why this is the case except that their developers simply didn't consider putting these parts in more accessible places. The same is true of fasteners. Even if you've standardized and reduced the number of fasteners in a product, if you don't make them accessible, the parts will likely not be separated or recycled. Making any metal fasteners magnetic can both ease disassembly and increase the likelihood that the fasteners themselves will be recovered for reuse or recycling.

> **Anyone who has ever tried to change the oil filter on a typical small car, or a fuse in just about any car, knows the pain involved when parts aren't easily accessible.**

Standardized components: Standardization can make components easier to replace and repair. It can help products be more easily used and understood, as well as upgraded. Most electronics would not be possible without a vast number of standards for everything from hard drive sizes to file formats to transistor connectors and power outlets. Modular components can extend this technique to make products more easily understood, used, serviced, repaired, and ultimately recycled.

No fasteners: Sometimes, cases and components can be designed to clip together without the need for fasteners like screws. For example, many mobile phone cases (like those for the Nokia 6200) do this to allow a plethora of third-party custom case designs. The Macintosh IIcx and IIci family excelled in this respect as well. Hard drives, fans, power supply, motherboard, and other components simply snapped into place with plastic tabs molded into the case itself. These bent just enough to allow them to be held aside for the components to be removed. Parts can be glued or bonded, but only where they are recyclable together because they are identical materials.

Part material labels: No matter how the parts are assembled or how easily disassembled, if the materials for each aren't immediately identifiable, they won't be recycled. Recyclers can't take many chances in contaminating their material streams. If they don't know that the part is a specific type of plastic or alloy of aluminium, for example, they won't throw it in the right bin. Instead, they'll usually divert it to the trash or to the shredder (where it may get contaminated even more, requiring it to be further downcycled). Each part should be clearly marked with an internationally understood label or icon declaring what it is. If some parts are just too tiny (like screws), they should all be made of the same material (so someone could safely assume that they're all the same material). If a part is made from an unexpected material or a material that looks like something else, this is even more critical. Metals that use alloys (like aluminum cases) should also be labeled by the alloy. It's not enough to just say "glass" or "aluminum" if it's a special kind that shouldn't be mixed with others.

> No matter how the parts are assembled or how easily disassembled, if the materials for each aren't immediately identifiable, they won't be recycled.

There are several commonly understood labels and indicators for just these purposes (see Figure 12.2). For example, the most commonly occurring plastics use a series of seven numbers within a recycling symbol (how's that for clear?). This system, though less common, extends to glass, metals, batteries, and other materials (see Table 12.1).

topic for sustainability club

FIGURE 12.2
You are probably familiar with a few of these material indicators, but the system is quite extensive.

TABLE 12.1

Recycling Symbol Key	
Plastics	
#1 PET	Polyethylene terephthalate
#2 PEHD or HDPE	High-density polyethylene
#3 PVC	Polyvinyl chloride
#4 PELD or LDPE	Low-density polyethylene
#5 PP	Polypropylene
#6 PS	Polystyrene
#7 O(ther)	All other plastics
#9 or #ABS	Acrylonitrile Butadiene Styrene: monitor/TV cases, coffee makers, cell phones, most computer plastic
Batteries	
#8 Lead	Lead-acid battery
#9 or #19 Alkaline	Alkaline battery
#10 NiCD	Nickel-cadmium battery
#11 NiMH	Nickel metal hydride battery
#12 Li	Lithium battery
#13 SO(Z)	Silver-oxide battery
#14 CZ	Zinc-carbon battery
Paper	
#20 C PAP (PCB)	Cardboard
#21 PAP	Other paper, mixed paper (magazines, mail)
#22 PAP	Paper
#23 PBD (PPB)	Paperboard: greeting cards, frozen food boxes, book covers
Metals	
#40 FE	Steel
#41 ALU	Aluminium
Organic Materials	
#50 FOR	Wood
#51 FOR	Cork (bottle toppers, place mats, construction material)
#60 COT	Cotton
#61 TEX	Jute
#62-69 TEX	Other textiles

Recycling Symbol Key (continued)	
Glass	
#70 GLS	Mixed glass container/multi-part container
#71 GLS	Clear glass
#72 GLS	Green glass
#73 GLS	Dark sort glass
#74 GLS	Light sort glass
#75 GLS	Light leaded glass (televisions, high-end electronics display glass)
#76 GLS	Leaded glass (older televisions, ashtrays, older beverage holders)
#77 GLS	Copper mixed/copper backed glass (electronics, LCD display heads, clocks/watches)
#78 GLS	Silver mixed/silver backed glass (mirrors, formal table settings)
#79 GLS	Gold mixed/gold backed glass (computer glass, formal table settings)

Of course, these need to be easy to find and read. Often, the label is molded right into the part (requiring no paint or appliqué), but these can be difficult to read if they aren't molded deep enough to cast a reasonable shadow. If the label can't be molded into the material itself (clothing would be an example), a label needs to be affixed in a convenient and clear way.

Indicate separation points: It's much easier (and faster, making it less costly) to separate parts if the edges are clearly indicated. This can be done by a change in color, texture, or finish, pronounced groove, or instructions (whether molded into the parts or applied later). It works for adjacent parts of similar or different materials.

Indicate disassembly sequence: Both for *repairability* and recycling, indicating the proper sequence to disassemble complex parts will increase the likelihood that products are repaired or recycled correctly and completely. For example, large Xerox office copiers have extremely complex mechanisms. Even a simple procedure, like clearing a paper jam, can be a nightmare to fix. For decades, these copiers have used colored parts and numbers to indicate which components to check first, how to open them correctly, and how to reset them to work properly again.

Reduce use of paint: Paint can often contaminate materials streams that flow through factories and recycling centers, requiring them to be diverted into separate substreams or discarded completely. This is because it contaminates the purity of the materials used so that they can't be effectively recycled. So, the less used, the better. The same goes for ink. This isn't always possible, of course.

All of the previous techniques make it easier to separate parts and disassemble products, and they do so by minimizing the time required. This further reduces the costs associated with recycling since time is often the most expensive component of disassembly—or assembly, for that matter.

Finally, to truly develop products for easy disassembly, designing them from the beginning to be assembled and disassembled easily is the best approach. It is during the conceptual phases or product and service development that wholly new approaches can result in break-through ideas that eliminate parts, affect manufacturing costs, specify materials in different ways, and conceive of challenges in new ways.

Close the Loop

The ideal sustainability strategy is to close input (resources) and output (wastes) streams so that nothing is wasted and everything is recycled. This strategy also means that nothing harmful will exit to the environment via air, water, land, and so on. To accomplish this goal, coordination between multiple players (suppliers, manufacturers, sources, and sometimes retail locations and even customers) is usually required. However, developers and organizations can plan policy, such as take-back programs, and redesign processes and specifications for manufacturing and service in order to come near to this ideal.

Create Take-Back Programs

One common approach for creating a more closed-loop system is to implement a product take-back program. Product take-back programs ensure that customers can take products and packaging back to the stores where they were purchased and that manufacturers will take these products back from retailers in order to recycle and reclaim whatever material they can. While this sounds easy, it requires a lot of coordination. Some materials, like packaging, can go straight into recycling programs, where and when they exist. But most products are complex enough that they require disassembly, and in some cases, special handling in order to retrieve as much useful material as possible. In addition, many components can be reused (such as ink cartridges and disposable cameras), but usually only by the original manufacturer. These products often need to find their way all the way back to where they were created originally or at least to an authorized repair, service, or recycling center.

Take-back programs ultimately pressure manufacturers to rethink their end-of-life strategies and the outcome of their products. For example, when Germany began requiring retailers and manufacturers to take back all packaging and correspondingly started taxing garbage more steeply, customers became adept at stripping the packaging and extraneous materials, right there in the store, and carrying only the

Germany)
Take-back
programs

product and necessary accessories and instructions home with them in their own reusable bags. This action saved them money since it cut down on their trash at home, and the upstream pressure on disposing of this trash influenced retailers and ultimately manufacturers to redesign their packaging to minimize material use. The system wasn't without its unintended consequences (in some forests, trash was dumped in the middle of the night), but overall the effect on packaging was extremely beneficial (and remains so to this day). In fact, the packaging redesign (and material savings) that was necessary under these conditions was duplicated in places even without the same taxes and laws.

> **Take-back programs ultimately pressure manufacturers to rethink the end-of-life strategies and outcomes of their products.**

Product take-back systems require integrated collection points, training for disassembling and identifying parts and materials, mechanisms for up- or down-cycling materials, reuse of components, separation of technical and biological nutrients, and safe disposal of anything left over.

In many countries, take-back programs are mandatory for certain products. These nations are driving the innovation for making this process efficient and effective—and all of us, ultimately, will benefit.

common theme

River and Lake Economies

Walter Stahel, the creator of the term *Cradle to Cradle*, describes two different loops in closed-loop systems that make-up what he calls a "Loop Economy" (see Figure 13.1). The first is the repair, reconditioning, and reuse of products and materials. This could include repairing a printer, refilling its ink cartridge, or giving it to someone else when its value to the original owner has passed. The second loop is in recycling the materials within a product after it finally wears out or is no longer usable.

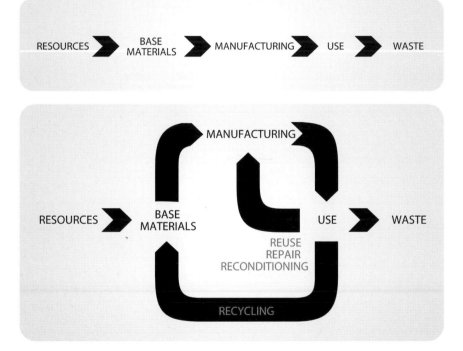

FIGURE 13.1

River economies (above) have materials, energy, and value run through communities, leaving little value behind. Lake economies (below) pool resources and value locally for longer lingering value and impact.

These loops are the first steps toward more efficient use of materials, but closed loops extend even further. As we saw in the LCA and Cradle to Cradle frameworks, waste represents a lot of valuable material and a lot of embodied energy. Reclaiming wastes through recycling or designing waste out of the final stages of use represents two ways to close this end of the loop. In addition, there are ways of minimizing the materials and energy involved with manufacturing by coordinating efforts between many different suppliers and partners and even co-locating them for greater effectiveness. This is called an *industrial estate.*

Industrial Estates

Industrial estates are gaining popularity in parts of Europe, but they take more planning and coordination than simply redesigning and building a factory. Consider the energy required to extract, smelt, cure, and refine ore into steel. Once poured into ingots, this steel is then cooled, transported to another supplier (often across vast distances), heated again to be formed into a new shape, cooled, shipped, heated, formed, cooled, and so on, repeating the process many times. Each time the steel is heated and cooled, a lot of energy is used and wasted. This is the case for many materials, including other metals, glass and ceramics, plastics, and so on.

Now consider a series of factories co-located so that the order can be heated and processed, and then moved to a final mold in one step, without cooling and reheating several times and without interim forms. Consider how co-located factories could quickly and easily move and transform wastes from one process into inputs for another. This is what an industrial estate (also known as *industrial symbiosis* or *industrial ecology*) is designed to enable. When designed, built, and run correctly, an industrial estate can realize enormous efficiencies over standard systems, while increasing recycled content (at least on the industrial side) and decreasing dissemination of toxic or undesired substances (since they're more localized, they're easier to manage). These are perfect solutions, but they require tighter and tighter closed-loop systems.

In his book, *The Performance Economy*, Stahel describes these loops in larger terms, adding financial and human capital to this model, and comparing them to traditional solutions. He contrasts the "River Economy," where materials, goods, services, and profits run through a community without leaving lasting benefit (especially economic benefit), to the "Lake Economy," where these materials pool in communities for greater local benefit.

Lake economics allow communities to retain capital in all its forms and amplify them in the form of interest, jobs, and assets. Lake economies benefit from closed-loop thinking and must deal with both the products and wastes of systems. In essence, it's in the community's interest to manage and value the entire life cycle and the rewards and responsibilities that come throughout.

> **Closed-loop systems offer opportunities for new companies to create new solutions and services that reclaim value in previously discarded (or buried) materials and previously unlinked flows of information material, energy, and services.**

Not everything, however, can be made in a lake economy. Sophisticated products and services that require rare materials or expertise can't be distributed everywhere. It's not likely that iPhones will be sourced, manufactured, sold, used, reclaimed, and recycled in thousands of towns across the U.S. Nor is it likely that every kind of fruit can be grown everywhere in every climate, or that every hospital will have experts in every treatment on staff. Regional differences in knowledge, materials, resources, and climate will always create the need for interlinked economies. However, the effects of small towns being drained of financial and other resources due to multinational corporations that move these resources, over time, to their headquarters can be countered through more thoughtful implementations of services in a lake-type economy.

Closed loop systems offer opportunities for new companies to create new solutions and services that reclaim value in previously discarded (or buried) materials and previously unlinked flows of information, material, energy, and services. Informationalization and transmaterialization (described in previous chapters) are two strategies that can enhance closed-loop systems and lake economies. Likewise, all of the frameworks described in Chapter 3 can be employed to identify opportunities to enhance efficiencies and find value where none exists today.

Kalundborg, Denmark

On the west coast of Denmark's Zealand Island is a town known as the best example of an industrial estate, Kalundborg, Denmark. Though first settled in the 1100s, in the beginning of the 1960s, a long, slow process of cooperation between the town government and businesses began with a pipeline to supply fresh water to a new refinery. Now, five core partners and many smaller companies interchange resources, material waste, and energy in an efficient web that has created over $120M in savings on investments of $60M (see Figure 13.2).

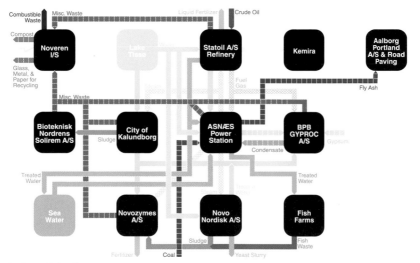

FIGURE 13.2

The Kalundborg Industrial Estate coordinates inputs and outputs from many area organizations in order to use energy, materials, and waste efficiently.

For example, the Statoil refinery consumes water from a nearby lake, crude oil, and excess steam from the Asnaes power station and outputs water and fuel gas to the power station, as well as sulfur (cleaned from emissions) to a company that creates sulfuric acid. The power station consumes coal and water (partly from the refinery) and outputs steam to heat homes in the city and to Novo Nordisk, a pharmaceutical factory responsible for 40 percent of the company's production of insulin, gypsum to Gyproc, a manufacturer of plasterboard, and fly ash to Aalborg Portland for paving roads.

Kalundborg, Denmark (continued)

There are other benefits as well (this is only a sample,) and while this model evolved naturally (it was not planned from the beginning but self-organized over time), there's no reason but lack of imagination and the will to act that prevents other communities from reaping even more benefits by planning industrial estates for themselves.

www.symbiosis.dk

William Good

In 2007, Goodwill Industries of San Francisco, San Mateo and Marin counties partnered with Nicholas Graham, founder of Joe Boxer, to create and launch a new business remanufacturing haute couture outfits from clothes donated to Goodwill (see Figure 13.3). The purpose of this new company was to choose clothes and materials from the massive amounts regularly donated and have designers make original pieces worthy of an expensive boutique.

FIGURE 13.3
William Good's re-creations are fun, stylish, and often surprising.

One of the reasons Goodwill is entering this market is that it simply has way too many clothes in its warehouses for the amount of demand available in their current stores. By crafting an offering for a different market, they hope to expand their ability to recycle, reuse, and resell the material donated to them, as well as train and employ people to give them usable skills for the future (their primary mission).

The William Good brand, and similar ones for other categories of donated goods, helps the organization close the loop between disposal and manufacturing. By taking in items customers are finished with, Goodwill creates a direct link for them to go back to manufacturing and new use.

www.shopwilliamgood.com

CHAPTER 14

Design for Effectiveness

E co-effectiveness is one of the main tenants of the Cradle to Cradle framework outlined in the book of the same name. Eco-effectiveness is the next step after eco-efficiency, and it seeks to go beyond simply increasing efficiency of materials and energy use with present technologies in order to create new solutions that create closed-loop systems, eliminate toxic material use, and erase trash. The idea behind eco-effectiveness is not only to eliminate waste but also to eliminate the concept of waste. You could say that this is the culmination of all the principles listed in this book working together.

> The idea behind eco-effectiveness is not only to eliminate waste but also to eliminate the concept of waste.

Extrapolate into the Future

For designers of all kinds, once our design processes return radically more efficient instances of current solutions, it's time to consider solutions that look at challenges in a completely new way and deliver more effective value. To do this, designers and developers need to reframe and reconsider what customers, organizations, and systems of all types need. Instead of reconsidering what the car, phone, kitchen, or office of the future might be, we need to question what transportation, communications, family, food, and work are all about and how these might change in the context of social, environmental, and market systems to be more sustainable and fulfilling. Instead of designing the next great mobile phone in the context of its relationship to the individual, her use needs, and the technological systems with which it must interoperate, eco-effective solutions examine the relationships that communication has with a myriad of social, financial, and environmental systems without presupposing that the solution will need to be a phone at all.

... designers and developers need to reframe and reconsider what customers, organizations, and systems of all types need.

The challenge of this type of approach is that the current paradigm of design and development projects—especially those done for external clients—doesn't allow designers to explode the project brief to such proportions. This kind of revolutionary change has to be strategic within an organization or system and despite the best intentions of most organizations, they just aren't thinking this strategically. In addition, most organizations' processes don't allow representation from designers, who often understand customers in exactly the ways and depth that others in the organization don't, or engineers, who often understand technologies and their potential more than others in leadership. This challenge calls for a new relationship between design/development and the rest of an organization— particularly in its strategic leadership. In addition, it requires additions or changes to the development process, not only of solutions but also of strategy. These are described, in detail, in Chapter 16, but practically, this is the only way to engage eco-effective innovation into existing organizations.

Effective solutions also require a different type of understanding. Not only are we called to rethink and reframe the need and solution, but we also need to reconsider the boundaries of what we consider the solution itself to be. An efficient solution may be a new product or service, but an effective solution requires a new system or solution across several systems in order to realize effective change. These solutions may involve, necessarily, service ecologies, education and training programs (for partners and employees as well as customers), new stakeholder partnerships, and awareness campaigns that establish new paradigms.

Rickshaw Bags

At the beginning of 2008, Mark Dwight and partners launched both a new company, Rickshaw Bags, and its first product—one of the most thoughtfully designed and sustainable messenger bags ever created (see Figure 14.1). Since then, they've added several other bags that innovate not only the features and options for bags, but also the materials and manufacturing process. In particular, the company's focus on Cradle to Cradle design principles has allowed it to create a unique bag with unparalleled sustainable qualities.

FIGURE 14.1
Rickshaw's sustainable bag.

Rickshaw's first bag, the Commuter Bag, is made in three pieces. The first two form a chassis and are sewn in a factory in China in a limited number of colors. The third piece is sewn in San Francisco, where the bags are assembled on-the-spot to fill orders only as they come in, reducing both inventory and returns and eliminating the unwanted, end-of-season bags that get destroyed or land-filled by other companies. This strategy also allows Rickshaw to customize the bags on demand, using fabrics supplied by customers and

Rickshaw Bags (continued)

customizing trim and product variables without requiring retooling or inventory of the main chassis. In addition to the mass-customization model and the sensitive sustainable design, the bags also excel in the kind of features people need, with innovations like Velcro silencers and magnetic closures, as well as optional attachments and modular inserts that can transform the standard bag for a knowledge-worker into the ideal baby bag to hold everything a busy parent needs. Most importantly, the bags are designed to be durable and last at least a decade, if not several. They're repairable and upgradable, and the beautiful flap can even be replaced if customers are tired of the fabric color or style.

The Commuter Bags are made from 100% post-consumer recycled PET bottles and are waterproof and PVC-free. The new Zero Bag line (in three sizes) has an ingenious design that results in zero waste and is made completely from nylon—both the soft fabric and the hard buckles. This means that the bags can be recycled with no disassembly since every piece is the same material.

Few products, let alone bags, have been as thoughtfully planned and manufactured as Rickshaw's. These are some of the best examples of how innovative, sustainable solutions can create better solutions for customers, company, and planet together.

www.rickshawbags.com

Restore

These days, it's not enough merely to reduce, reuse, and recycle. Impacts across the environmental, social, and even financial spectrum require designed systems and solutions to help restore our natural, cultural, and financial systems. Those designers and developers operating at the highest levels of design aren't just looking to make products, services, and events with lower impacts, they also want their impact to have a positive, beneficial effect on the world. For these designers, "zero impact" is not enough.

The Sustainability Helix framework is one of the few that makes room for this perspective. Restoration is one of the explicit states an organization can achieve (and the last one). The deeper descriptions within the helix include principles and strategies for organizations to move beyond mere leadership and describe how organizations can integrate these strategies across their operations and every aspect of their business function.

For designers, this requires some new thinking and reframing of the purpose of the solutions they create to include wider environmental, social, and financial goals. In her book, *Utopian Entrepreneur*, Brenda Laurel describes the use of "grand strategies" that pair a client or company's business strategies (which may or may not include sustainability of any kind) with those of the design and development team. Whether or not your project has been described in restorative terms (or even experimental ones), you have the ability to bring these strategies into the process.

In some cases, restoration can come from the materials we use. For example, there are many plants with restorative properties. (Actually, all plants have the potential to restore ground soil and air quality, but many will do far more than this.) Many plants are chosen because they can actively remove metals or toxins from the soil or air, as shown in Table 15.1.

TABLE 15.1

Plants with Restorative Properties	
Plant	Effect[*]
Alpine pennycress (Thalaspi caerulescens)	Accumulates cadmium and zinc
Indian mustard (Brassica junea)	Remediates selenium
Helianthus annuus	Filters heavy metals out of water
Polystichum acrostichoides	Reduces arsenic levels in soil
Aster frikartii[**]	Filters mercury
New England asters, joe-pye weeds, bluestems	Remediates PAHS (polycyclic aromatic hydrocarbons)
Poplars	Exudes proteins to attack environmental contaminants
Ferns	Remediates arsenic[***]

[*] Steven Rock, scientist at EPA's national Risk Management Research Lab in Cincinnati › ID
    Mag June 2003
[**] Engineered by Clayton Rough at Michigan State University
[***]Edenspace director of research, Michael Blaylock's work

Restoration can also be the focus of goals and processes employed, not merely the outcome. For example, involving locals in building their own communities can alleviate poverty and isolation better than simply bringing in outside workers temporarily. The old saying *"Give me a fish and I will eat for a day; teach me to fish and I will eat for a lifetime"* points to just this approach. In fact, in order to remediate social and financial impacts (as opposed to environmental impacts), process is probably the one most important factor. Decades of monetary and food aid around the world, for example, have done little to alleviate hunger and poverty and sometimes even worsened the problems.

Rethinking systems will be necessary to change conditions for the better, not merely creating less impact but also affecting change. This doesn't mean that we have to discard everything we know or have built, but it requires us to understand the challenges more deeply and approach them with an open mind about where the solutions might lie.

Restoring processes is just the first step in restoring whole systems (whether eco, financial, or social systems). This is where designers and developers, along with other stakeholders, can create the most change.

CHAPTER 15

Design for Systems

P erhaps the most revolutionary approach to designing sustainably is to consider the systems view and context of the things we design. As we learned in Chapter 1, everything we develop lives within an ecosystem of connected components, products, services, supply chains, and impacts. In fact, every solution lives within several ecosystems relating to environmental, sociocultural, and financial systems, as well as those for each of our stakeholders. Often, a re-examination of these systems reveals the most sustainable change possible, since it allows us to concentrate on the root of problems instead of merely on the symptoms that are most visible. Returning to the idea of systems solutions, it appears that redesigning platforms and re-engineering infrastructures will not only lead to more sustainable solutions but also to more holistic contexts for future solutions.

However, this theory puts designers into new territory. Not only are we usually inexperienced with supply chains, financial systems, and many cultural impacts, but the last thing our clients and companies want to hear when they engage us is that "we need to back up here and examine whether the whole system needs to be readdressed," or "this is really a cultural issue, and it's not solvable by simply making a new product."

Perhaps the most revolutionary approach to designing sustainably is to consider the systems view and context of the things we design.

Systems design calls for us to reinvent more than solutions; we also need to reinvent the *platforms* and *infrastructures* those solutions rest upon. For example, issues of traffic and car use are integral to the design of suburbs. We can't simply do away with cars or institute light rail because the interior structures of many communities don't support a system appropriate for urban settings in rural or suburban ones. Instead, we need to rethink and reframe the challenge in order to see new solutions. These system interconnections come to mind when evaluating solutions. For example, in choosing the most sustainable car, should cars be compared only to other cars, or should they also be compared to public transportation, like buses and subways, as well as personal transportation, like bicycles and scooters? A true systems view looks at the larger perspective in order to define needs and requirements at these higher levels before setting off to develop solutions that may only reinforce outdated systems.

> ... the last thing our clients and companies want to hear when they engage us is that "we need to back up here and examine whether the whole system needs to be readdressed" or "this is really a cultural issue, and it's not solvable by simply making a new product."

Designing a new car, for example, may do little to improve the situation even if it is small, light, fast, and efficient. A better solution might be to help more people move to other modes of transportation (such as bicycles if weather and distance permits, or casual and traditional carpools, or car-sharing services). Dense light rail might work to alleviate some of the challenge in combination with buses as well. If we think about this as a *transportation challenge*, rather than a *car problem*, we're more likely to create innovative solutions that actually create change. However, these solutions almost always require coordination among stakeholders and across larger systems in order to be built.

Curitiba, Brazil: An Example of Sustainable Innovation

Systems change is often the most difficult to affect since it involves so many goals and stakeholders. Often, the timing is critical for change to occur, and it always involves exemplary, and often visionary, leadership.

Such a time came to Curitiba, Brazil in 1972. The city had just elected a new major, Jamie Lerner, a trained architect. Mr. Lerner had already helped produce a new urban plan to ease traffic and urban sprawl. Curitiba was suffering under inadequate transportation for a burgeoning population and endemic poverty and slums. The city had also become increasingly inhospitable to pedestrians and citizens, and the rich social fabric of Brazilian society was unraveling. Lastly, it wasn't a rich city and, therefore, couldn't afford the expensive solutions that other larges cities could afford and put into place (such as an underground subway system).

One of the first moves that Mayor Lerner called for in the plan was to announce the closure of November 15 Street[1] to automobiles, making it a pedestrian walking street. The outcry from drivers and delivery people was immediate, and they planned to ignore the closure on its first day and "take back" their street through force. Fortunately, Mayor Lerner was a step ahead. When the cars arrived, they found the street filled with children playing. Cars never drove along these streets again. November 15 Street has become the city's favorite shopping and dining street. It's a popular place to meet and socialize, and the shops and restaurants that initially claimed they would be ruined by the closure have seen their businesses skyrocket and their location become one of the most desired in all the city.

This is often the case with systemic change. Those who do not see new possibilities cannot see or understand new solutions to newly uncovered opportunities.

Another innovation in Curitiba is its bus system (see Figure 15.1). The city did not have the funding to dig underground tunnels for a subway system. Instead, it instituted an improvised system using long, articulated buses and dedicated street lanes. This above-ground subway functioned much like any other subway in the world. There were designated stations that required tickets, and you could only enter a bus through one such station. These stations were built as above-ground glass tubes next to the dedicated bus lanes. Now, this above-ground subway system is so effective that 85 percent of Curitiba's population uses it. And it was built at a fraction of the cost other cities spend on subways. The system isn't perfect, and it required radical changes by car and truck drivers, but the overall solution has been better for most involved, including the poorest citizens who had no way to move around the city economically before. Other cities are now following this model worldwide.

---

[1]   In Latin America, it's popular to name streets after the dates of important national events.

FIGURE 15.1
The above-ground subway system.

From a cultural perspective, this became an important innovation as well. As Mayor Lerner has said, "A city with ghettos—ghettos of poor or of rich—isn't a city." The new transportation system lined different parts of the city's cultures and allowed them to interact in a new, healthier way.

This wasn't the only innovation that Mayor Lerner's administration started. They created small plots within the ghettos and subdivided city land to give to the poor, with a kit of building materials to self-build small homes, yards, and a tree. By no means were these large or luxurious, but it created homeowners out of squatters and engendered personal pride and ownership of vast portions of previous squalor. These plans converted areas prone to flooding into parks to boost the green belt and reduce financial and personal damage during bad weather. The city has since created free education centers (called "lighthouses of knowledge") to help educate those who can't afford it, train people for jobs, offer library services and Internet access, and help administer city welfare services.

www.curitiba.pr.gov.br

Leverage Points for Intervention

Developing at the systems level requires a strategic mindset that sees opportunities for change in intervention points. Donella Meadows understood this inherently when a realization occurred during a strategy meeting. She started listing leverage points that could be found in almost every system in order to create the most expedient change for the least amount of effort. Since then, she and others have refined this into a list of 12 points.

Leverage Points: Places to Intervene in a System, Donella Meadows, The Sustainability Institute, 1999 (in increasing effectiveness):

12. Constants, parameters, numbers (such as subsidies, taxes, standards)

11. The sizes of buffers and other stabilizing stocks

10. The structure of material stocks and flows (such as transportation networks, population age structures)

9. The lengths of delays, relative to the rate of system change

8. The strength of negative feedback loops, relative to the impacts they are trying to correct against

7. The gain around driving positive feedback loops

6. The structure of information flows (who does and does not have access to what kinds of information)

5. The rules of the system (such as incentives, punishments, and constraints)

4. The power to add, change, evolve, or self-organize system structure

3. The goals of the system

2. The mindset or paradigm out of which the system—its goals, power structure, rules, culture—arises

1. The power to transcend paradigms

All of these are places in which designers may find success creating larger system change. But translating these into products and service offerings isn't straight-forward and must be explored systematically and with multiple perspectives from knowledgeable partners in order to craft effective and lasting solutions.

Often, new business models await this type of systems thinking. When we look at traditional models in a new way or from a new perspective (often, corresponding to a different kind of stakeholder), we see new opportunities. New configurations or technologies allow us to shift resources from one system to another or from one product to a different service. New kinds of investments in human, natural, and financial capital can change product, service, and organizational efficiency, labor, component, or offering cost, life cycle impact, relationships with partners, retailers, and other stakeholders, and ultimately revenues and profits. New models can come unexpectedly from seemingly ridiculous frames of current challenges.

TerraCycle

In 2001, two Princeton students started TerraCycle with an unusual premise: they would collect garbage, sell it as garbage, use garbage for packaging—and make a profit doing so. Specifically, they collected food scraps, transformed the scraps into compost via worms, and sold it in reused plastic containers and cardboard boxes they collected from the trash. In addition, they decided to pay people for their garbage and still make a profit off the compost they sold. How's that for an innovative business model?

Tom Szaky and Jon Beyer began by collecting food waste from Princeton's dining halls, but since then, they have expanded considerably. They now sell their compost through Whole Foods, Home Depot, and Walmart stores. They also buy used bottles, bags, corks, juice boxes, and other trash from schools and online users to use as containers for their many products. They've expanded from fertilizer to bird food and feeders, liquid cleaners, deer repellant, rain barrels, bags and backpacks, and unique pots made from electronic waste (see Figure 15.2).

TerraCycle (continued)

FIGURE 15.2
Making garbage profitable.

By understanding economic, environmental, and waste systems and rethinking business models, they've found a valuable opportunity where others didn't—not to mention a solution.

www.terracycle.net

What doesn't work well in systems thinking, however, is trying to force people to adopt new models, understandings, or behaviors. We can educate, cajole, convince—even bribe—but forcing people into changes they aren't yet ready for is seldom successful. This includes trying to scare them into changing their behavior. Despite the wealth of assumptions to the contrary, people aren't often rational, and this makes markets equally irrational.

Instead, we need to make information available, help people (in all stakeholder groups) understand new possibilities, and give them the tools to make their own decisions. But we can't change their behaviors for them. This one fact creates an obvious drag on progress, and hampering it even further is the fact that usually only under severe circumstances (like emergencies or catastrophes) do people make such abrupt moves. This makes it all the more important to understand customer motivations and desires because these are the pivot points on which societies change decisions about themselves, their actions, and the world around them.

> **We can educate, cajole, convince—even bribe— but forcing people into changes they aren't yet ready for is seldom successful.**

Work from the Inside Out

One of the most successful ways to approach systems change is to start inside yourself and your organization. By understanding your own biases, preferences, values, and core meanings, you're better able to relate to others. Likewise, by understanding the mission, culture, goals, and vision of your organizations, you're better able to act on the values and core meanings that enable and support them. One good tool is to assess the values and core meanings for all key stakeholders, including customers, partners, competitors, and so on. This is especially true for the development team itself. When our values aren't correctly engaged and supported, we can't align our work to the biggest drivers within us. Finding overlap between customers, company, and team is most important. To whatever extent these attributes are different than those of our competitors, that creates opportunity to differentiate offerings. It's critical to find the parts of the system that will benefit from sustainable change in order to gather support for innovations that may feel foreign to many stakeholders.

Involving trusted stakeholder representatives in the process is also recommended. Each perspective within the system will offer new solutions to changing the system itself. For example, if you think that a material, component, or service supplier is the key fulcrum where systems change needs to occur, you need them involved if you hope to help that change take effect. In fact, the very act of involving stakeholders in cooperative development may create opportunity where lack of communication and understanding existed before. So often, two related parties don't discuss common concerns simply out of tradition and the fact that they've never done so in the past. Thus, opening these lines of discourse can enable development where none was possible before. It's common, for instance, that researchers never encounter customers, and therefore they can't direct their research toward the goals and strategies that organizations set simply because they don't know enough about the customers for whom they are creating.

Ultimately, designers and developers must find ways to question every aspect of "business as usual"—from materials to processes, to policies, to partners. Each time a piece of the system is addressed, there is an opportunity to innovate that piece into a more unified whole that accomplishes sustainable goals. Of course, this has to be done carefully and appropriately. When everything is in question, there often isn't enough information that is solid or has a decision attached to it in order to move forward. It's also seldom possible to change everything at once simply because of the vast details involved. However, over time, this kind of questioning can make the most powerful change occur.

> **Each time a piece of the system is addressed, there is an opportunity to innovate that piece into a more unified whole that accomplishes sustainable goals.**

Here are some things to consider reframing in your discussions in order to redesign systems:

- **Ownership** (of physical or intellectual property)

- **Consumption** (as a means toward creating or sustaining something)

- **Identity**

- **Motivation**

- **Understanding**

- **Leadership** (Where is it coming from, who is leading, and what are the effects? Who are the catalysts?)

- **Change** (Who and what does it enable or threaten and who is best prepared for it?)

- **Life Cycle** (Is it possible to eliminate, streamline, or redesign one of the steps in the life cycle: manufacturing, transportation, distribution, use, maintenance, recycling, disposal, reuse, and so on?)

- **Process** (Is it possible to eliminate, streamline, or redesign one of the steps in the development or maintenance process?)

A Note About Caution

The exact origin of the Precautionary Principle isn't known, but it dates back to 1800s common law and in more specific forms, to the 1930s in German law. Recently, it has been adopted by many government, NGO, and business organizations as a way of considering the systems implications of actions and policies. The European Commission adopted it in 2000 as a way of evaluating policies against unknown future outcomes.

The purpose of the Precautionary Principle is to recognize the need to anticipate risks and consequences before policies are enacted (or changed). This, inherently, acknowledges responsibilities on the part of those developing, approving, or implementing policy to be sure that these policies do not cause harm to people, society, or the environment. It can easily be translated to the development of products, services, events, and organizations and the actions within.

In the event of disagreement or lack of scientific consensus, the Precautionary Principle places the burden of proof on those proposing changes to systems or new actions to prove that these proposals won't cause harm.

Precautionary Principle: "When an activity raises threats of harm to human health or the environment, precautionary measures should be taken even if some cause and effect relationships are not fully established scientifically."

This raises questions for designers, developers, leaders, and policy-makers that we don't normally address, such as the following:

- What criteria should we use to evaluate new solutions, offerings, or technologies?

- Who has the right to say "no" to new solutions, offerings, or technologies?

- What are the right ways to say "yes" to a new solutions, offerings, or technologies?

- How can we improve our ability to predict the consequences of new solutions, offerings, or technologies?

Precautionary Principle: "When an activity raises threats of harm to human health or the environment, precautionary measures should be taken even if some cause and effect relationships are not fully established scientifically."

Busting the Segway

I'd like to be clear at the outset why I chose to represent the Segway as an example of poor systems design from a sustainability perspective (See Figure 15.3). I have enormous respect for Dean Kamen and his team of incredibly talented and dedicated engineers and designers.

Aside from the ridiculous hype surrounding the Segway before its introduction (basically calling it the most significant invention since fire) and the equally ridiculous reactions it generated (such as the city of San Francisco banning them before they were even available), the Segway is a wonderfully designed and engineered product (see Figure 15.3). Its interface design, industrial design, and mechanical engineering are all exemplary. It's an incredible innovation and thoroughly original. It even excels at aspects of sustainable design (like parts labeling, disassembly, dematerialization, and so on).

Its purpose, too, is intended to reduce the need to drive a car for solo travelers, which represents a significant reduction in material and energy use.

However, there's one problem: It doesn't serve a need. In fact, not only does it add no value that I can see, but it also actually potentially exacerbates existing social problems, such as obesity, since, in effect, it is a product for lazy people to be even lazier.

If the Segway is intended to replace larger transportation devices (such as cars) with larger impacts, it fails. Even if people aren't interested in walking instead of driving, using the same concept, other transportation options are still far better. Bicycles, scooters, skateboards, skates, etc., all provide the same function at far less impact in environmental, social, and financial terms.

FIGURE 15.3
The Segway is an example of fantastic innovation, just not fantastic sustainability.

Busting the Segway (continued)

We might assume that the Segway has value helping people with mobility problems (foot, knee, or hip injuries) get around easier, but tooling around standing isn't going to alleviate these problems, and these are better served by a smart wheelchair, like the excellent iBot, which Dean's team also created (it traverses stairs gracefully, runs across a myriad of surfaces, including sand, and stands up to put occupants at eye level with others). Those people with these kinds of mobility problems often also suffer from poor balance as a result of illnesses or injuries that cause the need for mobility aides, a condition the Segway would only exacerbate. In addition, these people are the ones who, most likely, still need a walker to navigate through the building, store, or home the Segway gets them to, once they leave it at the door.

So, despite the fantastic design process and development, the product doesn't have a compelling reason to exist. For example, it isn't competitive with other options in any of the five levels of significance. It doesn't offer performance or features that people are crying for, and it certainly doesn't deliver what it does offer at a price that's appealing to people (value). While it does connect with many people in emotional terms (mostly "cool"), it doesn't do this with any particular strength. Most Segway owners do connect at the level of identity, and it expresses for them a connection to the future, the joy of technological solutions, and these are clearly connected to the core meaning, wonder. However, it doesn't make these deep connections with many people because their other values (and sense of economic value) aren't engaged in comparison with other ways they can get to the store, office, or home.

This is why it's not sustainable (and hasn't been particularly successful in the marketplace either). Any needs it hopes to alleviate seem better solved in other ways (workers in Stockholm's airport, for example, simply use a Razor-like scooter at two percent of the financial, material, and energy costs). Any value it purports to provide seems misaligned with the market. The Segway solves a problem that doesn't exist, in a way that unnecessarily requires more materials and energy than other solutions. Ultimately, that's not good design, by any measure.

www.segway.com

Process

Designing radically more sustainable solutions doesn't take making a radical change to most development processes. In most cases, designers and developers simply need to actively ask questions about sustainability during the development process they already use. However, developers using older processes that don't investigate customer needs (like using ethnographic research techniques) and developers who don't develop customer requirements before they start developing technological ones will need to change their processes more substantially.

The additions to the development process aren't the only ones, however. To make truly sustainable products, services, and experiences, clients and companies need to rethink their priorities and strategies. This isn't a place where designers and developers are usually involved, but it is one that is critical to insert ourselves into. Too often, engineers and designers bemoan the fact that when they get project briefs, the description of what needs to be created doesn't fit their understanding of the customers or the market. At this point, it's often too late or too difficult to change corporate strategy, but developers often begin work on creating what they know is the wrong offering with the wrong set of features and performance, and a shallow understanding of what will be successful.

Traditionally, even the best development processes are focused on creating the best solutions possible (in the time frame available), given the parameters outlined by product marketing and corporate strategy. However, to be effective (in not only sustainable terms but also in market terms as well), organizations need to rethink what it is they should be offering in the first place. This realignment of the thought process is where innovation lies and where sustainability can have its highest impact, especially in terms of redesigning systems and reframing solutions, as discussed in the Chapter 18.

Developers aren't usually part of these conversations, but they often have the most accurate and influential data on what would satisfy and delight customers. Ethnographic research techniques, combined with more quantitative data, together build a much richer picture of customer needs, desires, and meanings than most strategic teams have access to. This new thought process is an invitation for developers to bring their deep knowledge into the strategic functions within an organization, although be forewarned that it's not always easy to do. There may be considerable reluctance on the part of your peers, such as marketers, who may feel threatened by an alternate perspective and description of customers, or from operations and finance who may assume that developers don't understand or appreciate financial, manufacturing, or other operational constraints.

A stakeholder perspective of the development process provides roles for everyone to play in sustainable development:

For **developers**, the process starts with learning about sustainability and making these goals visible within the development process. Often, these requirements will have to be driven from the development phases of the process and organization instead of the strategic phases (which is never ideal), and slowly migrated to strategic functions within an organization. This may take some time, but it is a necessity if the organization as a whole doesn't yet understand or value sustainability and its principles. Developers will need to educate their peers over time and insert themselves into the more strategic planning processes. However, in order to be effective, they will need to understand and appreciate these strategic processes, as well as the language, issues, and priorities of their peers in management and leadership positions within their own organizations and those of their clients.

For **leaders**, sustainability offers opportunities not only to differentiate brand and offerings, but also to increase any number of operational efficiencies, as well as to reap the benefits of attracting and keeping talented **employees** who feel an emotional and meaningful connection between themselves and their work. Sustainable advancements throughout an organization offer differentiable solutions not only to consumers but also to customers of all types, including governments, organizations, and other businesses. Supply chains are hardly ever under the control of one organization, so sustainability as a strategy almost always involves coordination with suppliers and partners.

For **customers**, more sustainable solutions offer enhanced functions and the potential to satisfy values through owning, supporting, and using these new offerings. In addition, more sustainable solutions can make it much easier for people to make better choices without becoming experts in sustainability themselves (see Chapters 17 and 18).

For **communities**, the rewards involve longer-term involvement between an organization and those it serves. This aspect might mean better relationships that lead to better cooperation, more resilient actions, more stable incomes, less waste and environmental damage, and more stable jobs.

Finally, for **investors**, sustainable organizations can be sources of more stable and lasting growth and return on investment. In addition, supporting organizations that also support the same values generate deeper and more satisfying emotions, values, and meanings for customers.

Innovating Solutions

There are several different development processes currently in use around the world. Depending on the industry and product or service, developers may use any of the following:

- **"Waterfall:"** A step-by-step, phased approach that finishes one phase before starting another. This often results in slower development but more complete specifications and more successful interfaces and solutions. This development, however, can lead to disconnection and compartmentalization of teams and inefficient use of some resources while prior phases are being completed.

- **"Extreme" Programming, Spiral Development, and "Agile:"** A concurrent developmental approach, especially for software, that starts technical development immediately and rapidly iterates it. Unfortunately, many agile developers try to iterate both customer requirements and technical requirements (front-end and back-end) concurrently as well, usually resulting in interfaces that inadequately meet customer needs. Because of the pace of development, decisions are often not always strategically driven but instead are tactically driven. This development approach requires close communication and coordination among the team and constant team-wide check-ins.

- **Skunkworks:** An approach to separating a team from the "normal" developmental process and the rest of the organization, in order to innovate where innovation has been difficult otherwise. This works best with multidisciplinary team members representing all aspects of development, including marketing, operations, manufacturing, human resources, service, and so on.[1, 2] One drawback of a skunkworks approach is that once the team project is completed, the team's organizational learning is often dissipated when its members are reintegrated back into the rest of the company. Also, if this is the only

---

[1] "Skunkworks" was a term coined by engineers at Lockheed in the mid-1940s to describe their secret ad-hoc group developing advanced military aircraft.

[2] The original Ford Taurus was developed using a skunkworks model. The dedicated team, representing all divisions of the company, was separated from the rest of Ford in order to develop the Taurus.

way for a company to innovate, it says something significant and troubling about the organization's present culture.

- **"Genius Design:"** This is a common approach that is comfortable for designers, engineers, and marketers alike (because it doesn't rely on customer research), but it is the most risky and least successful (because it doesn't rely on a deeper understanding of customers' needs discovered in customer research). The most famous and successful practitioner of this approach is Apple, Inc., but without a leader like Steve Jobs (with his uncanny and unusual sense of what customers want and need), it would have been a failure (as most organizations who develop this way eventually discover). While it can be highly efficient and fast, it often leads to overly technical solutions that don't match what customers want.

- **User-Centric or User Experience:** Modern user/customer-driven processes make a point to research customer needs before solutions are specified. They use a variety of quantitative and qualitative research techniques and tools to integrate and express what is found via customer profiles and scenarios. They make use of user testing throughout the development process (before technical specs are decided and not merely for troubleshooting interfaces after the solution is mostly completed). It is often criticized for being more expensive and time-consuming for development, and even with substantial time spent understanding users, the trouble in identifying and translating customer needs—and the support for doing so at leadership levels in an organization—can still lead to solutions that miss important aspects of the customer experience.

- **Six Sigma:** This is a specific, detailed approach to developing quality solutions created by Motorola in the late 1980s. Its purpose was to devise a reliable way to measure and increase quality throughout the development process. While it has seen gains in manufacturing and other parts of operations, it's not suited to strategic innovation, concept-generation, or understanding the customer experience.

- **Hybrid Models:** Most organizations use a mix of techniques from many of the previous processes in order to develop in a way that works for their culture. Obviously, some are more successful than others, and there's probably no one "right" way to develop anything. However, none of the previous techniques is inherently sufficient for incorporating sustainability easily into an organization. While it is possible to integrate all of the principles and strategies in this book into most processes, I'll illustrate this point using what I consider a more complete development process, because it incorporates both strategic and tactical development.

The Strategic Innovation Process

For the purposes of this book, I will be referring to a development process for strategic innovation, currently rising in popularity and combining many of the best elements of those from the past. Through it, it's easy to point to where sustainable criteria can best intersect the development of sustainable strategies, as well as products and services.

The qualities of this process are shown in two groupings: the development of corporate strategy (deciding what the organization should produce and offer) and the development of offerings (how best to create and deliver the offerings that meet an organization's strategic goals). See Figure 16.1.

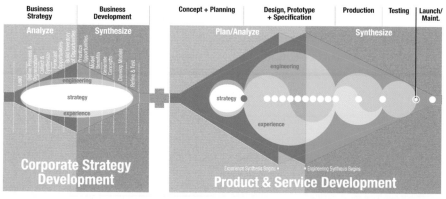

FIGURE 16.1
Strategic innovation requires involvement between both strategy and development.

It's important to note that it is the combination of these two phases that is most powerful to integrate sustainability into an organization. Doing so only in the strategy portion (and failing to communicate these strategies throughout the rest of the company) results in offerings that are not only less sustainable (in all aspects) but also have products, services, and events that aren't aligned with the purposes and direction of the organization. Likewise, implementing sustainability principles only within the development process may yield slightly more sustainable offerings but will miss opportunities to truly innovate solutions that shift paradigms and create more competitive, differentiable, and sustainable systems.

Both of these phases also include an analysis period in the beginning (shown in Figure 16.1 in blue) and a synthesis period in the last half (shown in the figure in green). While the exact activities in each may differ, there is a similar process of investigating context and opportunities (analysis) and then determining those best suited to the organization, strategy, and offerings (the synthesis). The synthesis portion of the strategy phase ends with the building and articulating of an organization's strategy. In the development phase, the result is the building of the actual offerings, whether product, service, event, or some combination of all of these.

Note that there is a slight shift between the point at which analysis turns to synthesis in the development phase. This is because it's critical that the front-end (user experience) synthesis begin before the back-end (technical synthesis does). This allows decisions about the customer experience to drive technical implementation decisions, instead of the opposite (which is the source of many inadequate solutions that come to market).

Some of the aspects of the other processes inherent in this one include the following:

- **Waterfall:** While strategy is always ongoing within an organization, corporate strategies must be decided, defined, and communicated *before* offerings are developed if these offerings are to meet these objectives. Also, while engineering (or technical) research development can start almost immediately, key decisions about these aspects of the solution shouldn't be made *until* the customer research and front-end development (the *experience*) is decided in order to ensure that the back-end will adequately support the intended experience. Similarly, some production and testing of well-understood components can begin early, but it's usually *less* efficient to perform the majority of these steps *until* the final solution is defined and built. Otherwise, much of the production and testing must be redone every time significant changes are made.

- **"Extreme" Programming, Spiral Development, and "Agile:"** Within each of the subphases, "extreme" techniques are very helpful. The tight integration of teams, communication, and rapid prototyping often lead to fast development that is informed by all involved. The trick to making this truly effective (and not merely efficient) is being sure that the desire for fast results and the ability to iterate don't cause technical decisions to pre-empt customer experience decisions on the grounds that the work can be fixed later (it won't be), or that it's too late since those aspects of the engineering have already been completed (which only ensures that the engineering department will efficiently create an unsuccessful offering).

- **Skunkworks:** It's always important to involve all aspects of an organization's activities in the development of offerings, particularly in terms of sustainability. Even if representatives from all departments aren't on the development team fulltime, regular check-ins and feedback will help ensure that what seems like an innovative solution in one aspect doesn't create insurmountable problems in another. Also, separating a multidisciplinary team from the rest of the organization in order to concentrate on development—particularly if the corporation's culture doesn't support innovation well—can often increase both efficiency and effectiveness in terms of innovating systems-oriented offerings, but it's critical to specifically incorporate ways of documenting and sharing the learning within the team so that it can be shared effectively with the rest of the organization. This learning is at least as important to an organization as the final product or service—if not more so.

- **"Genius Design:"** It never hurts to have sharp people on a project—it's preferable, in fact. Some team members have natural talents in looking at a complex challenge and seemingly pulling miracle solutions out of thin air. This is an asset but one that must still be validated along the way, both with customer and operational requirements—not to mention from a lens of sustainability.

Integrating the best of all of these processes isn't easy, but this is the most effective way to work for many organizations. Now that I've described the various parts of the process in general terms (and most readers should see many similarities with the processes they already use and experience), we can address how to integrate sustainability strategies and techniques into these steps.

The Strategy Phase

In some ways, this is the most important phase in sustainable development because it is here that the most change can be made, both in terms of what an organization creates and offers as well as in how it informs and motivates the rest of the organization to integrate sustainable principles in every aspect of its operations. See Figure 16.2.

FIGURE 16.2
Innovation strategy involves analysis that widens possibilities and synthesis that narrows them to the best strategies.

Multidisciplinary Perspectives

It's important to approach strategy from multidisciplinary perspectives. This is the biggest mistake that most leaders make. They feel that only they will truly understand the organization's opportunities or that this information is so critical that it can't be entrusted to many in the organization. The result is that leaders unintentionally misinform themselves about market, customer, and their own corporation's requirements, and they fail to involve and inform the rest of the organization in details critical to supporting the decided strategies. So, as with the development process, it's important to involve others in the organization that may know the requirements better and may see opportunities that leaders don't. Stakeholders outside the organization can be key contributors as well. Strategic alliances with suppliers and partners often yield sustainable, competitive, and differentiable strategies. Likewise, "participatory design" involves representative customers on the development team to help ensure that offerings create value for them.

It's critical to include customer or user research during this phase because this represents a deeper understanding of customers than traditional market research uncovers. This is the most common mistake leaders make when investigating and formulating strategy, and it represents the first point at which sustainability criteria intersects the development process.

> It's important to approach strategy from multidisciplinary perspectives. This is the biggest mistake that most leaders make.

Often, support for sustainability in all its forms (social, environmental, and financial), at both the corporate and customer levels, is lacking or invisible because traditional, qualitative market research techniques aren't adequate at measuring or uncovering aspects of customers' needs and desires at the levels of emotions, values, and meanings. These are where sustainable values intersect the customer experience. So it's easy for quantitative market research to show that the most important driver of customer decisions is price (because it's most easily measured). Even qualitative data show, time and time again, that customers regularly spend more—whether they intend to or not—when their values or emotions are triggered by the product or service options.

Sustainable values, too, are triggered by offerings, and these triggers change depending on the customer segment and category of product or service. If an organization has no way to discover these, it has no way to appreciate them when making corporate strategies. This is exactly where designers can have the most impact in an organization (either their own or for a client). The designer's customer research techniques are particularly good at uncovering evidence of emotions, values, and meanings, and designers are particularly adept at communicating often ethereal or esoteric information. To be sure, designers need to understand their audience (leaders and managers) as well as they do their users, in order to successfully communicate this information in a way that fits their culture, but it is often the only access to this critical understanding that these strategy-makers ever get.

Strategic Partnerships

Conversations at this level about sustainability can open the door for strategic partnerships with other organizations, both for mutual support and for more successful implementation. Sometimes, this helps minimize risks of retaliation or ridicule (such as from issue-oriented NGOs and watchdogs). Other times, this can add or help develop important, sometimes expensive, new tools, materials, processes, or other solutions in order to speed the realization of better practices and offerings. Seemingly unlikely alliances are now commonplace (such as between Home Depot and the Forest Stewardship Council, or McDonald's and PETA, People for the Ethical Treatment of Animals). These alliances aren't simply "cover" or validation for companies to dress themselves in more acceptable guise. Most issue-oriented NGOs already have considerable expertise in their domains that organizations would spend too much time and money trying to duplicate otherwise. These alliances can be critical to speed the effectiveness in developing, implementing, and maintaining a sustainable agenda.

Understanding via Examples

During the strategy-development process, it's important to present
scenarios and even proofs-of-concept that can embody opportunities
in visual and visceral ways so that business leaders can understand and
envision what they might be like. Here, too, designers and other developers
can play an important role. While these examples are strictly prototypes
(since they don't actually work), they may be rigged to look quite real
and work in limited ways, mimicking the key behaviors that would make
them successful if created. In the development of services, it's common
to show the artifacts of the service through different points in the process,
even if these are simple screens or paper output (such as a receipt showing
a purchase or other transaction). Because services are so intangible in
comparison with products, this is sometimes the only way to help leaders
and managers envision an innovation, especially the more literally minded
of them.

> ... it's important to present scenarios and even
> proofs-of-concept that can embody opportunities
> in a visual and visceral way so that business lead-
> ers can understand and envision what they might
> be like.

Sustainability principles and criteria need to be inserted into the research
component during strategic analysis. Specifically, these contexts need to be
actively investigated—often, in order to be seen at all. It needs to be part
of the investigation into market contexts (such as competitive and industry
research), customer contexts (actions, behaviors, needs, and desires), and
operational contexts within the organization (technologies, capabilities,
efficiencies, and so on).

Part of the Vision

Lastly, sustainability is strongest when it is part of the vision and mission
of an organization. This might require a reexamination of the purpose of

an organization and a conversation with all relevant stakeholders, including investors, owners, and employees. It's true that sustainable strategies are often driven by motivated leaders, but even the most forceful leaders can't drive these values alone. They will need to communicate effectively and seek out key people within (and sometimes outside of) the organization who can help communicate and enact these values throughout the corporation.

Efficiency

Where the culture of an organization isn't quite ready to discuss sustainability purely in terms of environmental and social values (it's likely that most organizations are already prepared to have a discussion about financial sustainability), there are several proxies that can stand in for them while they develop within the culture. For example, sustainability often aligns with the drive for efficiency, whether in terms of cost-reduction of enhancing performance and productivity. This creates a great opportunity to create a sustainability agenda under the very real guise of efficiency and effectiveness.

Brand Differentiation

Another approach might be on product or corporate brand differentiation. While there are no accepted standards for actually valuing brand in financial terms (there are several frameworks offered by different advertising, brand management, and consulting corporations, however), most leaders and managers do already accept that brand differentiation *does* have value in forming a more meaningful connection with customers. Even if this can't be measured well, many executives will accept sustainability as a principle if it can be shown that customers or industry will respond in brand recognition and admiration.

> **Where the culture of an organization isn't quite ready to discuss sustainability purely in terms of environmental and social values ... there are several proxies that can stand in for them while they develop within the culture.**

Risk Mitigation

Still another valuable strategy for bringing sustainability into an organization is in terms of risk mitigation. Even when leaders and other stakeholder don't understand or accept what they feel are "touchy-feely" values, most can imagine the costs associated with operational risk in policies and offerings. For example, many manufacturers have discovered the negative impact ill-defined or non-existent social policies with their labor can have on their brand and revenues. Even when a corporation's policies are deemed acceptable to the public, if their products are created and serviced by subcontractors with unacceptable labor policies, they still risk the backlash. Risks not only affect revenues, but also can affect any aspect of an organization's operations—and many, many environmental and social issues defined in the first chapters of this book.

Divesting in Apartheid

In the 1980s, when corporations and university investments were found to support apartheid in South Africa, they suffered a very real divestment risk when this became an issue with the public. These organizations had to change policies quickly, shift investments, and even sell business units in some cases, to mitigate the risk of losing investments, short- and long-term revenues, customers, and their market value.

These kinds of social and environmental risks are difficult to measure, which can be an ally to those promoting more sustainable practices and strategies within an organization. The open-ended potential—especially in financial terms—of these risks, piled on top of each other, can concern leaders and shareholders beyond what the actual risks may be. This is further leverage to make change in favor of socially, environmentally, and financially sustainable practices. However, using this approach to scare or threaten action usually has the opposite effect. Many companies, when risks are uncovered under threat, move quickly to squash any discussion or knowledge of these risks *precisely* to limit their liability if these risks ever result in legal action. Threatening to "go public" with risky behavior can result in stifling all conversation of change and all progress toward more suitable and sustainable policies.

The most helpful frameworks to employ in the strategic phase are the Natural Capital and Sustainability Helix frameworks. This is because they are both high-level, business-centric discussions of sustainability. Both tend to support strategic goals and speak in ways that leaders and senior managers can understand within their own cultures. This doesn't mean that the others aren't useful. All of the others, particularly Natural Step and Cradle to Cradle, can offer important and vivid examples that help strategists better understand sustainability and envision how it might benefit their organizations. But these frameworks don't offer a lot of operational tools for strategists to use to implement sustainability throughout their organizations, particularly in strategic terms.

The Development Process

Once organizational strategies are already aligned with sustainable goals, sustainable development becomes much easier since every decision about better environmental, social, and financial performance is already supported at the highest levels of the organization. The very products, services, and events (and the infrastructure and platforms that support them) will already start to be reshaped in favor of sustainable goals. For example, if the strategies are set correctly, the projects that head for development should not only be more sustainable solutions (perhaps, focusing on cleaning rather than a particular cleaning product), but they should also automatically support the organization's efforts to become more sustainable in a unified way. See Figure 16.3.

> Once organizational strategies are already aligned with sustainable goals, sustainable development becomes much easier since every decision about better environmental, social, and financial performance is already supported at the highest levels of the organization.

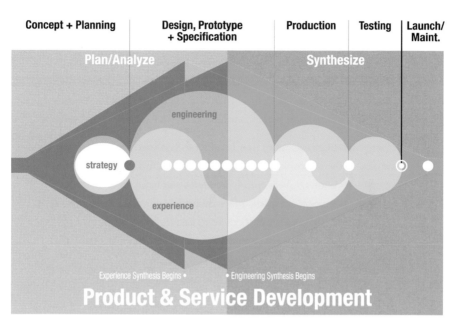

| Concept + Planning | Design, Prototype + Specification | Production | Testing | Launch/ Maint. |

FIGURE 16.3

As with the strategic phase, the development phase involves an analysis phase of widening possibilities and a synthesis phase in which decisions are made regarding the solution.

However, even if the company's strategies aren't yet infused with sustainable goals and more appropriate offerings, the development phase offers the opportunity to make these offerings, whatever they are, more socially, environmentally, and financially sustainable. Finding a sympathetic and supportive ally among an organization or client's leadership can be critical as well. The more support you have making decisions, the more likely your solutions are to be chosen and implemented.

Using any of (or a combination of) the frameworks outlined in Chapter 3, development teams can infuse sustainable values into their work alongside whatever the other values expressed by the organization may be.

Biomimicry, in particular, should be used in the conceptual and prototyping phases of the back-end or technical solution (engineering). Looking for analogs and lessons in nature will have a huge impact on the technical solution. While it can be an important source of inspiration for the customer experience solution—particularly in appearance—because nature doesn't share many of society's values, it may have less of an impact here.

Frameworks of evaluation, like Life Cycle Analysis, may be difficult to employ in the conceptual and earliest prototype phases since there are so many unknowns. *Very* rough estimates can be made, and the further along the development is, the more that can be measured. However, other frameworks, such as Datchefski's Total Beauty, may be more helpful in developing innovative concepts and solutions in these early phases.

To some extent, teams may have leeway to redefine criteria and requirements. Usually, the details of a product or service's implementation are sufficiently loosely defined that there is a lot of room to interpret and develop a sustainable agenda. These may be justified to managers under the guise of issues outlined previously (such as risk mitigation, efficiency, brand differentiation, customer loyalty and connection, and so on), or they may go unnoticed by senior management altogether. Whether acknowledged and appreciated or not, the goal is to reduce environmental, social, and financial impacts nonetheless, and designers, engineers, and other developers have a lot of influence on achieving these reductions since the development of the solution is almost entirely in their hands.

This is the phase where all of the strategies described in Chapters 4 through 16 come to bear. Each one of these should be considered in the conceptual parts of development, as well as in prototyping and testing. The entire team should be acquainted with these strategies, even shallowly, and it would help to use a checklist, like the one in Appendix A, to check against progress throughout the development process.

In order to move effectively and efficiently through development, the strategies more important to address in each subphase are shown in Table 16.1.

TABLE 16.1

Most Useful Frameworks and Strategies by Subphase		
Subphase	**Frameworks**	**Strategies**
Concept and Planning	Natural Capitalism	Localization
	Sustainability Helix	Transmaterialization
	Cradle to Cradle	Informationalization
		Design for Effectiveness
		Design for Systems
Design, Prototyping, and Specification	Total Beauty	Design for Use
	Cradle to Cradle	Dematerialization
	Biomimicry	Substitution
	Natural Step™	Design for Efficiency
	Natural Capitalism	Localization
	SROI (Social Return on Investment)	Transmaterialization
		Informationalization
		Design for Durability
		Design for Reuse
		Design for Disassembly
		Close the Loop
		Design for Effectiveness
		Design for Systems

Most Useful Frameworks and Strategies by Subphase (continued)		
Production	LCA (Life Cycle Analysis) SROI (Social Return on Investment) Cradle to Cradle Total Beauty	Dematerialization Substitution Design for Efficiency Localization Transmaterialization Informationalization Design for Durability Design for Reuse Design for Disassembly Close the Loop Design for Effectiveness Design for Systems
Testing	LCA (Life Cycle Analysis) SROI (Social Return on Investment) Cradle to Cradle Total Beauty	Design for Durability Design for Reuse Design for Disassembly Close the Loop Design for Effectiveness Design for Systems
Launch and Support	LCA (Life Cycle Analysis) SROI (Social Return on Investment)	Design for Use

All of these strategies and frameworks should be viewed as a set of potential tools. Not all of them will be useable in all projects and contexts. Some teams will have expertise and find it easier to use and implement some over others. At the very least, they should be considered periodically by the development team, at least at the start of each subphase, in order to address whether they can add value to the process. If they can't add value at the moment or on a particular project, then perhaps they can be used on the next project or at the next phase.

At the very least, development teams should represent the needs of the market (marketing), the customer (design), and the operation (engineering), as well as those responsible for managing the development (project management). This is the minimum list to develop, and it defines skills and priorities as much as job titles or departments. Expanded teams might include representatives from manufacturing, law, finance, customer service, engineering specialties, material science, biology, customers, partners, community, and so on, but the basic set of four outlined here is critical. Without representation of the needs and requirements from these four areas, a project runs the risk of failing any one of these important requirements.

Any team member should be allowed—and encouraged—to offer suggestions across the entire project. Sometimes, someone *not* immersed in the day-to-day details or assumptions of a discipline has the clearest perspective on innovative solutions. This doesn't, however, mean that everyone is responsible for every decision. A clear distinction should be made between sources of ideas and responsibilities for deliverables. The former should be open to everyone willing to participate. The latter must be clearly defined and communicated, both in terms of authority and in terms of schedule. Every decision and deliverable must be someone's responsibility to make (though, hopefully, not before adequate and open opportunity for fresh approaches and exploration).

> **Any team member should be allowed—and encouraged—to offer suggestions across the entire project.**

One way developers can approach these issues with clients or within their own companies is to begin with evaluations and analyses of current products, services, and events in order to raise the awareness about these issues and set a baseline for improvements. This might be done officially or unofficially, paid or not. The benefit of these evaluations is that they educate all involved, and they create the opportunity to discuss specifics instead of generalities. They should be approached as a "stake in the ground," in order to calibrate solutions rather than as a threat for further action. In addition, using the Sustainability Helix to evaluate the organization itself can help elevate these issues beyond the offering level and into the strategic level. These evaluations can be rough to begin with, but the expectations of their process, measure, and accuracy should be set accordingly in order to eliminate misconceptions about progress or intent.

Despite the potentially dire need for new solutions to make a revolutionary process in terms of minimizing social, environmental, and financial impacts sustainably, there are no perfect solutions and any gains are important gains. Most projects will fall far short of the desired and needed gains. It's easy for development teams to focus on what they weren't able to accomplish instead of the gains that were made. This isn't to say that the focus must be on inspiring *big* change and exponential gains. Now, more than ever before, we need vast advancements in all areas of sustainability (reference the first two chapters again), and these only come with sustained focus and inspirational—and aspirational—leadership. But the world doesn't change overnight, and turning and moving in the right direction a little is preferable to continuing in the wrong direction.

> **Most projects will fall far short of the desired and needed gains. It's easy for development teams to focus on what they weren't able to accomplish instead of the gains that were made.**

Measuring Results

A s we learned in the discussion of frameworks in Chapter 3 (especially for LCA and SROI), not only are there no standards of measuring sustainability progress, but also where there are mechanisms in development, the data and process are quite difficult to decipher (as with a deep Life Cycle Assessment).

This leaves developers at a loss as to how to weigh alternatives, and it leaves customers at a loss as to how to choose between them. In addition, it makes it difficult for organizations to track their progress.

Where the best measures exist so far is in the Socially Responsible Investing (SRI) sector. SRI funds have been around since the 1970s, and they measure publically-traded companies on their activities and investments in order to choose those whose values fit the fund investors'. As such, there are *many* different funds that measure and rate companies along a different set of values.

For example, some of the earliest SRI funds started as a way for religious investors to put their money to work in the service of their religious values. These funds might not invest in companies that support gambling, alcohol, or tobacco in their products and services or in their own investments. These funds exist in many religious communities, although the largest SRI funds have no particular religious affiliation.

The Calvert family of funds (with over 30 different varieties) represents some of the oldest SRI funds, and it screens companies on the basis of issues such as labor policy, corporate governance and transparency, product safety, human rights, weapons manufacturing, and so on. Calvert doesn't invest in companies that don't meet these baseline criteria. On top of these, however, it offers funds that focus on a variety of other criteria, such as alternative energy and global development. Other well-known funds from organizations like Highwater,[1] Domini, Parnassus, and Morningstar offer different mixes of criteria. The growth in this investment approach has even spawned a backlash (though a small one). The Vice Fund purposely

---

[1] See Highwater's criteria on page 24.

invests in everything that most SRIs won't invest in, such as gambling, pornography, weapons, and so on.[2]

The point of this discussion is to show the variety of issues and measurements used by SRIs (the most sophisticated measurers so far).[3] These funds compile their list of investment companies in two ways: the first is to *screen* companies for certain behaviors. These might be positive screens (such as having an equal opportunity hiring program) or negative ones (such as investing in or owning nuclear power plants). These screens are binary: companies are either in or out based on whether they pass the screen. The second mechanism these funds use are ratings on various other criteria. Companies might get a high, medium, or low, or an SRI might apply a point system that reflects the degree of activity for a particular criteria.[4]

In the end, all of these funds develop a list of approved companies to invest in, and only then do these funds research the financial performance of those left on the list to decide which to include in their portfolio. (SRI funds are still looking for the biggest financial returns they can get, just not from companies that actively work against their values.)

A last issue to acknowledge in the SRI investment world is one of the enduring controversies. Some investors don't ever want to invest in a company doing "bad" things (from their perspective). However, others acknowledge the influence they can have by investing in (and, thereby, rewarding) those companies that are making the most changes in their industry. This reward, they feel, sends a more powerful message to the entire industry to change rather than alienating them altogether. For example, many sustainably-minded investors would never want their money to be invested in McDonald's because they feel it is the epitome of an unsustainable, unhealthy business model. (As a result, McDonald's doesn't show up in many SRI funds.) Others, however, feel that by

---

2   VICEX:US vicefund.com

3   www.socialinvest.org/resources/mfpc/screening.cfm

4   www.sri-advisor.com/funds/analyzer.cgi

investing in McDonald's, a company that has changed more than probably any of its competitors in terms of environmental, human rights, and animal rights issues, they are rewarding the very behavior they want to see and creating pressure for other companies to follow suit.

The SRI approach illustrates a possible path for organizations to develop their own mechanism for measuring progress. They can set positive and negative screens for everything from product or service category (safety, usability, meaning, etc.) to material and energy use to partner strategies. Each organization needs to create its own set of screens, based on its on values, however. There are also many other things that can be measured and rated, such as material and energy impact, toxicity, quality, durability, recyclability, and so on. They can even adopt the criteria (screens, ratings, and all) from some of the most popular SRI funds as a start. For public companies, it would be smart to assess which funds they already qualify for (or not).

It would be great if someone created a standard that we could all refer to and measure against. It's probably not realistic, however, at least for the social criteria. Life cycle assessments are an attempt to do this for the material and energy flows throughout a product or service's life cycle, but because values are so different for individuals (not to mention organizations), SRIs may be the closest we can get to standards on the social side.

There are a few attempts to create scorecards for different subsets of these issues. Some are industry specific (like the Pharos measurement tool for building products), while others try to be more universal.

Most of the existing popular scorecards were developed before sustainability issues became as popular as they are today. Some, like the Balanced Scorecard, developed in the 1990s, integrate some human capital issues with the traditional financial capital issues. But none of these is complete.

Global Reporting Framework

One of the most popular sustainability measures, at least for social and governance issues in corporate policy, is the Global Reporting Framework.[5]

---

5 www.globalreporting.org

This standard, defined by the Global Reporting Initiative, is one of the few widely-recognized frameworks for reporting corporate-level policies. Covive, a San Francisco-based design firm, has created an excellent summary of this framework that clarifies many pages of documentation.[6] (See Figure 3.23 on page 99)

The GRI Framework lists criteria in six categories, such as the following:

- **Environmental** (materials, energy, water, biodiversity, emissions, effluents and waste, products and services, compliance, and transport)

- **Human Rights** (investment and procurement practices, nondiscrimination, freedom of association and collective bargaining, child labor, forced and compulsory labor, security practices, and indigenous rights)

- **Labor Practices and Decent Work** (employment, labor/management relations, occupational health and safety, training and education, and diversity and equal opportunity)

- **Society** (community, corruption, public policy, anti-competitive behavior, and compliance)

- **Product Responsibility** (customer health and safety, product and service labeling, marketing communications, customer privacy, and compliance)

- **Economics** (economic performance, market presence, and indirect economic impacts)

Balanced Scorecard

In 1993, Robert S. Kaplan, a professor at Harvard, and David P. Norton created the Balanced Scorecard approach to measuring corporate performance. Beginning with the publishing of their book, six years later, this became a rapidly adopted and popular approach. In particular, their approach was designed to measure financial (and some human) capital

---

6 www.covive.com/gri

performance against corporate strategies and mission. This was one of the early attempts to align corporate strategies with corporate missions in a way that could be measured numerically and continuously in terms of dollars.

The Balanced Scorecard[7] measures performance in four main areas:

- **Financial** (Are the organization's strategies and tactics creating financial value?)

  Cash Flow, ROI, ROE, etc.

- **Customer** (Do the organization's offerings provide value to customers?)

  Customer satisfaction, loyalty, retention, etc.

- **Internal Business Processes** (Are the organization's processes driving value in the first two areas?)

  Accident ratios, effectiveness, activities, etc.

- **Learning and Growth** (Is the organization growing in terms of human, information, and organization capital?)

  Investments and returns, employee turnover, gender and racial ratios, etc.

Walmart Packaging Scorecard

Walmart, as part of its much-publicized shifts to explore and implement sustainability, is developing scorecards of its own in order to rate products and services it sells and the suppliers and partners it works with. It first focuses exclusively on packaging. Like similar forays into sustainability metrics, it quickly found that this seemingly "simple" approach is actually quite complex, and there are few models to follow. The promise, however, is this: Since Walmart is the biggest company in the world, it can have the biggest impact on the most organizations and for the most customers. That's a good reason to persevere.

---

7 www.balancedscorecard.org

Walmart Packaging Scorecard (continued)

Walmart's packaging scorecard is still developing, so its criteria are somewhat in flux. No doubt learning what works and what doesn't over time will improve it, and the best lessons will come from actual implementation, something that hasn't started yet in earnest. So far, the scorecard categories reflect metrics and equations for a variety of environmental criteria:[8]

- 15% Greenhouse gas emissions (GHG/CO_2 per ton of production)

- 15% Material value

- 15% Product to package ratio

- 10% Cube utilization

- 10% Average distance to transport materials

- 10% Recycled content

- 10% Recovery value

- 5% Renewable energy

- 5% Innovation

Walmart hasn't published the metrics within these categories, but the ratios give an idea of what it finds important in sustainable packaging. As this standard evolves and influences suppliers, it has the potential to impact all packaging, not only that destined for a Walmart store. If your company is already a supplier to Walmart (or hopes to be), this is already a standard that is important to you.

---

[8] walmartstores.com/FactsNews/NewsRoom/6039.aspx

Other Ratings and Metrics

There is a lot of hope for creating far-reaching standards for rating and labeling of products and services along sustainability criteria. Some standards focus only on the environmental aspects, some only on the social and governance criteria, some attempt to comingle them, and others hope to allow customers to mix their own ratios. For all of the interest in this area, the development is slow-going, partly due to the inherent complexity and partly because, although it seems that everyone wants to use these systems, it's decidedly not sexy in comparison to investments in "clean energy" and other sustainability innovations.

Some of the existing systems created for consumers include the following:

- **alonovo.com** (one of the oldest, having started in 2005)

- **goodguide.com**

- **betterworldshopper.com** (one of the most complete but only a book, not an online resource)

- **ethiscore.org**

- **hrc.org/cei** (Corporate Equity Index)

- **pharosproject.net** (focused on building materials)

- **www.c2ccertified.com (**Cradle to Cradle certificate program—very extensive and expensive)

Other rating systems are already in use (such as those developed by the various SRIs), but intended only for use by other businesses (making them expensive):

- **www.kld.com** (KLD Research)

- **www.innovest.com (**Innovest)

Some proposed systems, in development, include the following:

- **SBAR** (Sustainable Business Rating System)[9]

---

9 makower.typepad.com/joel_makower/2005/05/sustainable_bus.html

Some manufacturers aren't waiting for standards to emerge and are creating their own, for use with their products:

*   Timberland

*   Home Depot

*   Carbon Trusts/Tesco Carbon Label

All of these systems have their strengths and weaknesses. Some are focused only on single issues while others attempt a full-spectrum rating. It's safe to say that all describe various subsets of the total measurement solution.

The systems listed previously are ratings systems, meaning they rate companies (and sometimes individual products and services) in a way that allows them to be compared to others. These can lead to Type III labels (as defined by the ISO 14000 standards). They differ from binary seals (Type I) that only report whether a company, product, or service passed the screen. There are *many* of these in use throughout the world and across the spectrum of issues. Organic, Fairtrade, Kosher, and Dolphin Safe labels, for example, are Type I labels. A product with one of these labels designates that it meets the requirements of the certifying body, but products without these labels don't necessarily signal that they don't meet the standards. (The manufacturer may simply not have paid for the certification procedure.) In addition, it's not possible to compare two products, both certified by one of these binary labels, in terms of their relative performance in this area. While Type I labels can be important, they are ultimately less helpful in enabling customers to make choices based on their values and concerns.

All of these systems have their strengths and weaknesses. Some are focused only on single issues while others attempt a full-spectrum rating. It's probably safe to say that all describe various subsets of the total measurement solution.

According to Terrachoice Environmental marketing study, more than half the eco labels today hype some narrow claim at the expense of larger sustainability issues.

Label Types

Type I labels are product seals licensed by governments or third-party private entities based on a multitude of criteria or impact. For example, the U.S.-based Green Seal or Sweden's Nordic Swan Type I seals can vary substantially in their criteria, which may or may not be known or understood by customers.

Type II labels are informative, self-declared seals about the environmental qualities of a product, such as "contains 75 percent recycled paper."

Type III labels offer quantified product information based on a life cycle assessment. These labels are best for comparisons between products or services. There are few examples of Type III labels in use. One in development is the Reveal label.

Type IV labels are single-issue seals licensed by companies or organizations. Examples include the Leaping Bunny (signifying no animal testing), the Good Housekeeping Seal of Approval, Underwriter's Laboratories insignia, and the Forest Stewardship Council seal.

See examples of all of these types of labels in Figure 17.1

Type I Labels

Type II Labels

"Dolphin Safe"   "Organic"

Type III Labels

Type IV Labels

FIGURE 17.1
Because socially- and ecologically-oriented labeling address many different types of issues, it's difficult for customers to know which criteria in each are actually being measured.

Reveal Rating System

I've learned a lot about rating and labeling systems working on my own solution. Beginning in 2002, I started designing label solutions for making the complex criteria involved across the sustainability spectra clear for consumers in a shopping environment. My aim was to make it easy to compare products and services in the same categories, while still being informative and imparting a sense of validity to the data presented. Over the next four years, I developed a series of labels to do just this—the Reveal Rating system (named by a colleague at school), as shown in Figure 17.2. In addition, while at business school, I used this project as my thesis and quickly found that the label portion of the design was just the beginning of the solution.

FIGURE 17.2
The Reveal system is designed to grow over time (from the Phase 1 seal to the Phase 2 seal) as LCA (Life Cycle Analysis) data becomes available.

More critical (and complex) was the system behind the label that collected, stored, and rated the data necessary to track performance across so many criteria. The system is, necessarily, vastly more complex than what appears on the label, but it is where the real design challenge lies. In particular, I found that the sources of data were critical, as well as who stored and validated it, and which parts were made available for use. The system itself is what needs to be designed, and it needs to balance needs and concerns among a variety of stakeholders in order to be effective and to address those very same stakeholders' concerns as a prerequisite for participation.

The system needs to be available to business, organizations, and individuals in a variety of ways. For example, there is a lot of excitement for mobile phone applications that allow shoppers to call up ratings while in the grocery aisle. However, my research showed that, while exceedingly modern and cool, most shoppers simply weren't going to do this. Kiosks can provide another important interface for some shoppers, but the best solution was to get the labels right on the products themselves (alongside nutrition facts and other information). This is probably too much to hope for anytime soon—at least across a wide range of manufacturers. The next best interface, however, is to get the labels onto the store shelves, right next to the price tags. In order to be effective, any system attempting to provide this information must work across this variety of interfaces in order to offer shoppers the best solution for each person or context.

But this was still only part of the solution. In order to make this kind of system a reality, the next step was to develop a business model around it so that it could be financially self-sustaining. One of the issues here was that most start-ups require significant investment, and investors almost always want huge returns in a quick time frame (three to five years). Otherwise, they'll put their money in other promising investments (like clean energy technology). In addition, investors want a certain amount of liquidity, which usually comes from an IPO (Initial Public Offering)—something that probably isn't well suited to this kind of organization—or a buyout from a large corporation. (This is something that would likely invalidate ratings from the service since there would be inherent conflicts of interest for that company's products and services, as well as its competitors.)

These are some of the complexities around measurement systems that have, so far, prevented standards from emerging. The returns, however, are staggering. Probably more than any other innovation, rating and labeling systems can influence the most change faster than any other solution. The SROI calculations for merely three products, over 10 years, show billions of dollars in influence.

For now, designers and developers are on their own until standards emerge. Some existing standards can be used as a starting point, but most will fall short of some critical criteria concerning an organization's interests. Overcoming these challenges will require time and effort, but these solutions will pay off as organizations use metrics they develop to increase their operational, market, and financial performance over time. Not having a system of metrics only ensures that sustainable performance won't be measured and most likely won't be addressed.

Some of the challenges for creating a custom metric include the following:

- All stakeholders need to be aligned in mission and values (as closely as possible).

- Corporate strategies must support sustainability and be measured against sustainable gains.

- Not all performance can (or should) be measured in dollars. Some may need to stand on their own merits—such as social issues.

- Financial metrics must be integrated with social and environmental metrics. Keeping them separate keeps decisions about each (and their inter-influence) separate as well.

- There is no perfect system so metrics should be introduced, in beta form, quickly to speed organizational learning that will improve the metrics themselves. Waiting until a system is complete is going to be a long wait.

- Experts can help develop these systems more quickly if their participation isn't threatening to the organizations.

- Organizations need to put aside competitive pressures at this level and work together to establish more complete standards quickly. Companies can compete over performance according to the standards later, but until there is a standard, there's no way to compete effectively. The standard itself can't be an organization's competitive strategy.

- Funding for metrics must become a priority, as well as implementation and education throughout the organization.

CHAPTER 18

Declaring Results

*I*f results are so difficult to measure, how can we begin to declare them?

This is an important question, and it's at the heart of those charging organizations with "greenwashing." As we've seen, the issues are complex and often counterintuitive. In order to make reasonable claims, it's important to articulate benefits carefully and clearly within the context of the audience we address.

There are many different ways of describing the market for sustainable products, services, brands, and messaging. For sure, this market is growing, if only due to rising concerns over efficiency and health care. The number of people who respond to environmental and social issues is still quite small, however. Depending on how this market is measured—and by whom—it's still well under 20 percent of the total consumer public. Some of the most popular segments include LOHAS (Lifestyles of Health and Sustainability),[1] The Green Gauge Report,[2] Conscious Consumers, and the Cultural Creatives. Each is backed by a separate body of research, and for the most part each describes a very similar segmentation of customers. It depends to whom you are speaking.

LOHAS

The LOHAS segment comes from an industry group by the same name. It segments consumers into five groups, based on the degree of their commitment to sustainability issues when purchasing products and services:

- 17% LOHAS (the most committed consumers)

- 21% Naturalites (concerned most of the time, especially for natural and organic foods, but less involved with other practices)

- 19% Drifters (concerned occasionally, with good intentions, but price and other factors regularly outweigh environmental concerns)

---

[1]   Natural Marketing Institute, 2008

[2]   2007 Green Gauge Report

- 20% Conventionals (mostly not concerned but curious, and involved with some practices, such as recycling)

- 21% Unconcerned (not concerned at all or actively against sustainability)

Green Gauge

Another segmentation, Green Gauge, comes from GfK Roper Consulting, and it describes five groups, based on their buying behavior around environmental concerns (see Figure 18.1). This is probably the longest-running survey of "green" perceptions, and it has shown a lot of change just in the past three years (following very little change in the proceeding decade).

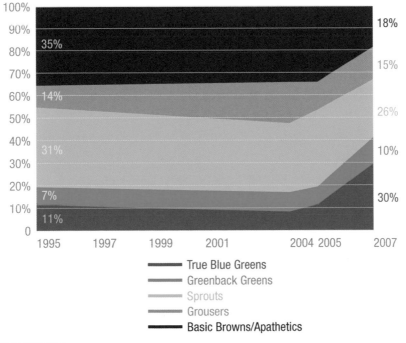

FIGURE 18.1
The Green Gauge market segments.

2007

30% True Blue Greens (most committed)

10% Greenback Greens (interested but not always willing to spend more)

26% Sprouts (undecided about environmental issues)

15% Grousers (view environmental issues as too big or complicated to do anything about)

18% Apathetics (not interested in environmental issues)

Where the most committed consumers (dubbed *True Blue Greens*) consistently accounted for only around 10 percent of the population in the U.S. for many years, beginning in 2004, it began to increase and jumped significantly to 30 percent just last year (2007). In total, the top three segments, representing consumers who respond to environmental issues to some extent has grown from 55 percent to 65 percent. These figures differ greatly from the results of other studies, however (see below).

2005

11% True Blue Greens (most committed)

8% Greenback Greens (interested but not always willing to spend more)

33% Sprouts (undecided about environmental issues)

14% Grousers (view environmental issues as too big or complicated to do anything about)

33% Apathetics (not interested in environmental issues)

2004

9% True Blue Greens (most committed)

6% Greenback Greens (interested but not always willing to spend more)

31% Sprouts (undecided about environmental issues)

19% Grousers (view environmental issues as too big or complicated to do anything about)

33% Basic Browns (not interested in environmental issues)

1995

11% True Blue Greens (most committed)

7% Greenback Greens (interested but not always willing to spend more)

31% Sprouts (undecided about environmental issues)

14% Grousers (view environmental issues as too big or complicated to do anything about)

35% Basic Browns (not interested in environmental issues)

In both of these segmentations, environmental issues were the focus. While social issues may have been included as part of this research, these segments didn't deeply research a broad range of social issues other than those dealing with the environment.

Conscious Consumers

BBMG provides us with yet another segmentation about the buying concerns of U.S. consumers, but it focuses more on overall social values and not merely those of environmental concerns. This difference starts to provide a more complete picture of buying values across a wider spectrum of issues. Like the two cited previously, their four segments represent varying degrees of behavior and concerns around environmental issues. For example, the top segment last year (2007) represents only 10 percent of consumers, instead of the 30 percent quoted by the Green Gauge Report:

- **10%** Enlighteneds (who make a point to reward companies along social and environmental goals)

- **20%** Aspirationals (balance their values with convenience and price)

- **30%** Practicals (prioritize price, quality, and efficiency over their social agenda)

- **40%** Indifferents (prioritize everything over any social agenda)[3]

---

3 BBMG.com

The BBMG study found that five values drive conscious consumerism: health and safety, honesty, convenience, relationships, and "doing good." The top two Conscious Consumer segments compared to the top two Green Gauge segments differ by 10 percent of the U.S. population (which is a lot). The Conscious Consumers research is much more similar to the LOHAS figures.

Cultural Creatives

1996

24% Cultural Creatives[4]

47% Moderns

29% Traditionals

One of the most important aspects of this study, which is shared by several others, is that socially- and environmentally-driven buying doesn't follow traditional demographic segments. People who buy according to their values come from all age, gender, income, and social groups. They aren't just students or "liberals" or "new agers" or "baby boomers." They may have diametrically opposed religious and political affiliations, in fact. What unite them are interests in comfort, health, personal growth, innovation, and balance, whether it applies to their homes, food, transportation, communities, or relationships.

These interests represent important lessons that anyone speaking to these groups should understand. Traditional approaches to reaching environmentally- and socially-minded customers probably won't be effective. The typical verbal and visual choices to signal and trigger meaning aren't effective as these segments grow to include more people of varying cultural backgrounds.

---

4 www.culturalcreatives.org

What Is Marketing?

Much to the delight of marketers that turn their attention to the top segments of all of the market segmentation studies listed earlier, the groups who respond most to sustainability also typically represent the most affluent and educated consumers—the ones marketers of all kinds treasure the most. These groups also represent the consumers who are most receptive to products, services, and events with improved social and environmental performance, so designers and developers should seek to understand these customers well—from social and behavioral perspectives, not merely demographic perspectives.

This requires us to explore market research techniques more deeply and to supplement them with our own user- or design-research techniques. In particular, quantitative techniques rarely illuminate details of social and environmental values. In order to understand customers in these ways, it's necessary to use ethnographic techniques, which are often more difficult, more time-consuming, and more ambiguous.

Designers, engineers, and other developers are often suspect of market data, partly because they don't get to see where it comes from (and how it was generated and summarized) and partly because it doesn't align often with what developers already understand about customers. For sure, many designers and developers firmly keep themselves in a little world, separate from the reality of markets that don't cater to their wants and desires. However, many are keenly aware of the world around them and constantly filter messages intended for many audiences. It is these designers that often have a better "read" on the culture of their customers, even if they lack the figures to validate these impressions.

> Marketing is the inhale. Sales, promotion, advertising, and PR are the exhale.

One of the most common misunderstandings about marketing is that it is most often combined with sales, advertising, PR (public relations), and other forms of promotions and messaging. Lumped together, this means that true marketing is often never undertaken within a company, or if it is, it's done shallowly. A better way to think of marketing is as the inhale process an organization does in order to learn from customers, competitors, the markets, and industries. It is the complement to the exhale representing what organizations attempt to message to these constituents (often at the expense of understanding who they are addressing). Organizations can't exhale effectively without inhaling effectively, which is one of the reasons so many traditional companies and industries are finding innovation so difficult.

Most marketing plans erroneously focus on the exhale at the expense of the inhale. Most marketing departments don't even know how to inhale very well. Sticking to traditional market research techniques (like surveys, polls, focus groups, and mall intercepts), where they do attempt to inhale customer insight, they usually focus on shallow, quantitative methods in an attempt to "go wide" and include as many customers as possible, but at the expense of "deeper" understanding of customers that can only be obtained through ethnographic techniques. This is another place in the process where designers' skills can benefit an organization because design research techniques favor ethnography, and they have tools to integrate what is found into actionable learning that can inform innovative development of products, services, and new kinds of solutions.

Most marketing plans erroneously focus on the exhale at the expense of the inhale.

This isn't to say that traditional techniques have no value. The best research comes from a combination of quantitative and qualitative research methods. Ideally, initial qualitative research will uncover important factors that can be explored more widely with smart quantitative research. The results of this research can be used in even deeper qualitative investigations that uncover important aspects of these factors. The process is never-ending because patterns and people shift over time, but moving back and forth between ethnographic and traditional techniques lets each inform the other, yielding better insights overall.

For forays into sustainability and describing benefits to social, environmental, and financial values, it's imperative that everyone on the development team truly understands not only what their customers need, desire, and feel, but how they express these feelings and why. Emotions, values, and meanings transcend price and features and often negate their value when customers become excited and engaged on these higher levels.

The best research comes from a combination of quantitative and qualitative research methods.

The way to understand this mechanism is to observe it first and then experiment with it. This experimentation is the only way to become familiar enough with how customers take in messaging relating to their values. Only then can organizations manipulate performance and message to work effectively together.

What to Say (and Not)

Because not all customers are socially and environmentally engaged—at least as their primary motivators (see the market segmentations above)— messaging around these values isn't always the most effective approach. It depends deeply on the exact customers you're speaking with, but in general, messaging around claims, benefits, and insights regarding *efficiency*, *health*, and *safety* are more successful propositions. It's not that customers value these benefits more than social and environmental benefits (although the majority do), but that these issues appeal to a wider group of customers that spans both those who are sustainability-engaged and those who aren't.

If a marketing strategy (with associated and supporting messages) emphasizes efficiency or health benefits in order to drive customers to a more sustainable solution, that's perfectly fine. For customers who respond to socially- and environmentally-targeted appeals (who number many fewer), these benefits serve only to reinforce their willingness to choose these products. For sure, these benefits should be mentioned, but they don't have to be the focus of the story in order to reach all of these customers.

In general, the attributes most valued in purchases of all kinds include the following:

- Convenience

- Efficiency and value

- Health and safety

- Performance and effectiveness (sometimes, this relates to features)

- Price (though value can be subjective and relative)

- Status and identity (which are all culturally-mediated)

- Honesty, credibility, and authenticity

- Social justice

Messaging to any or all of these is possible, depending on what customer research has determined are the best prospects for gathering customer attention and communicating benefits effectively. In addition, the same solutions can be messaged differently to different customer segments in different contexts (whether these divide geographically, demographically, by media, or any other attribute). As long as the claims aren't conflicting, it's fine to tout the deeper social and environmental benefits of an offering to those who will respond accordingly while touting health or efficiency benefits of the same solution to others in different contexts. Not all customers are ready to hear about all benefits, but for most, social and environmental performance isn't a turn-off for products they value for other reasons. Often, it's just another benefit, whether they value it or not.

In order to make successful claims about benefits—especially those involving social and environmental impacts—stick to honest, detailed descriptions in clear, plain language. Customer research around triggers will help you determine the right imagery, language, and approach, and periodic testing with customers can tell you when you're on or off target. A great guide to making claims successfully comes from the Australian Competition and Consumer Commission, with the following maxims:

In order to make successful claims about benefits—especially those involving social and environmental impacts—stick to honest, detailed descriptions in clear, plain language.

- Be honest, truthful, and accurate.

- Detail the specific part of the product or process the benefit refers to.

- Use plain language that an average member of the public can understand.

- Explain the significance of benefits (without being vague or overstating claims).

- Be able to substantiate benefits.

- Make claims only for "real" benefits (not suspect, though true, benefits that pale in comparison to bigger negative impacts).

- Pictures as well as text must be valid.

- Be sure claims clearly refer to packaging or contents.

- Claims should consider the whole product life cycle.

- Claims using endorsement or certification should be used with caution (as well as permission).

- Claims should not overstate the level of scientific acceptance.

- Avoid problematic, vague terms like "green" and "environmentally safe," "energy efficient," "recyclable," "carbon neutral," "renewable," or "green energy."[5]

Organizations need to scrutinize carefully their messages around products to be sure that they're not only accurate but also reasonable. For example, when McDonald's launched a line of salads as a healthier alternative to the hamburgers they usually offered, they didn't check to see if both the calorie count and fat content were actually higher than the sandwiches they were meant to replace (due to the salad dressings included). The company was rightly criticized for completely misunderstanding the point of healthier food.

However, many claims of greenwashing aren't fair either. When General Electric pledged a $4.5B investment over three years toward developing alternative energy technologies, some criticized the move as self-serving since the advancements would reap rewards for the company in terms of efficiency and revenues in expanded markets. However, that's exactly the point of innovation. Alternative energy development isn't supposed to be a charity. It's already a sound business practice. This fact is what's fueling the interest in sustainable practices among traditional businesses and industries. It's not a problem that companies are investing in ways that promise to create profits—that's what companies do. As long as these developments are more sustainable, they still represent sustainable progress.

These aren't the only guidelines that developers should follow, but they're a fantastic start. Companies shouldn't be afraid to make accurate, relevant claims if they do so appropriately. Likewise, they shouldn't be afraid of dissent or disagreement if it is reasonable and creates a health dialogue about the issues. For example, a famous 2006 discussion ensued between Michael Pollan, the author of *The Omnivore's Dilemma*, and John Mackey,

---

5   Green marketing and the Trade Practices Act 2008, Australian Competition and Consumer Commission. www.accc.gov.au/content/index.phtml/itemId/815763

the CEO of Whole Foods, after Pollan criticized Whole Foods in his book. Initially, Mackey wrote an open letter to Pollan on his company blog.[6] Then Pollan posted a reply on his Web site.[7] Almost a year later, the two met on stage at UC Berkeley for what was expected to be a heated debate but became more of a mutually respected discussion.

> ... [companies] shouldn't be afraid of dissent or disagreement if it is reasonable and creates a healthy dialogue about the issues.

More important than the points either made in the discussion was the fact that the discussion took place at all, and those watching it (it's available online as well[8]) were able to learn about the complexity of these subjects, not to mention where each opinion stands. That this discussion took place at all is a great sign that some of the players in the sustainability world (or, more accurately in this case, the organic food industry) are willing to discuss important aspects of these issues instead of merely beating each other up in the typically sensationalistic media. The result was undoubtedly not only good for both Whole Foods and Pollan but also for customers, the industry, and these movements.

It's important for organizations of all kinds to pay attention to the details of their actions and to be sure that their processes and policies (as well as their product and service offerings) align with their corporate strategies, mission, and values. Discussions like these can help calibrate when they are aligned (or not), and preferably these discussions should happen inside the company during development and not wait until offerings, actions, and messaging are finally launched.

---

[6] wholefoodsmarket.com/socialmedia/ jmackey/2006/05/26/an-open-letter-to-michael-pollan

[7] www.michaelpollan.com/article.php?id=80

[8] http://webcast.berkeley.edu/event_details.php?webcastid=19147

It's always a good idea for organizations to offer more details about their products, services, and policies, and their Web sites are a great place to do this. Since sustainability issues aren't simple, it's likely that messages— especially advertising—are going to be correspondingly simplified (but still accurate). Offering substantiating material about claims, as well as outlining assumptions, is one of the ways organizations can set context, explain their approaches, and dispel charges of greenwashing. Education about these issues expands exponentially as it is shared, and it helps create a more fertile foundation for future improvements in sustainable practices. This expansion extends to information about what to do with products once customers are finished with them, how to recycle them (and where), and how best to dispose of them (if necessary). Relevant information should be made available to all stakeholders (customers, vendors, clients, employees, partners, suppliers, communities, NGOs, governments, and so on) in order to facilitate the most progress (and earn the most respect).

> **Offering substantiating material about claims, as well as outlining assumptions, is one of the ways organizations can set context, explain their approaches, and dispel charges of greenwashing.**

When companies in industries that just can't be sustainable promote sustainable advances, it strains credibility, even if the advancements are real and accurate. For example, Walmart's and other retailers' current model of selling almost all of its products manufactured in China may never be sustainable. In addition, weapons manufacturers and energy companies whose power comes predominantly from coal or nuclear plants may be able to improve somewhat but still never be able to claim being a sustainable business in the larger sense. A slightly better coal-firing process can make important improvements, but these improvements pale in comparison to the damage they still do. Consider an example of a company that promoted the fact that they no longer used child labor for manufacturing on weekends. For most customers, the 29 percent reduction in child labor wasn't significant and didn't justify the percentage of manufacturing still dependent on children. Common sense is a perfectly good guide here, as is testing the claims with people beforehand. Too often, advertisers do their best to find *something* to say that's accurate and positive without stopping to consider how ludicrous it might still be from the public's point of view. It's better to say nothing or wait until an advancement is more significant than to test the ire of customers, communities, and other stakeholders prematurely.

When companies in industries that just can't be sustainable promote sustainable advances, it strains credibility, even if the advancements are real and accurate.

Lastly, it's unfortunately common for many companies to give money to a charity and then spend many times that amount promoting the fact that they did just that. This can be an easy check against greenwashing or acting against stated values for organizations. If you're spending more time promoting your good deed and advancements than you are advancing your values, you're definitely in the danger zone. Likewise, if your spending on lobbying against your values exceeds the amount that you are spending in favor of your values or goals, then you're not only in the danger zone for public messaging, but your actions are completely misaligned with your values. For example, in 2007, Ford, Toyota, and many other companies lobbied heavily against increasing the CAFE (Corporate Average Fuel Economy) standards while simultaneously pledging to raise the efficiency for their offerings and touting the benefits they'd already attained with hybrid models. In addition, Ford was showing off its newly renovated River Rouge Factory (its oldest and biggest), having been redesigned by McDonough Associates (one of the authors behind the Cradle to Cradle framework). The factory is now a stunning example of sustainability principles at work in the industry, but the juxtaposition of the two policies revealed both a lack of commitment to sustainability, as well as a badly convoluted and contradictory corporate strategy.

One of the most successful rebranding efforts to date, in terms of sustainability, is what BP launched in 2002. Under the leadership of CEO John Browne, British Petroleum hired Landor to rethink its brand identity in response to new strategies he was implementing inside the company. In order to communicate its new vision as a contemporary energy company with an eye toward the future, and differentiate it from other oil companies, the rebranding effort dumped the old name in favor of just the initials and promoted several new phrases to fit them, the most popular being "beyond petroleum." The addition of a new logo (dubbed the "Heilos") and a savvy advertising campaign created an unprecedented shift in consumer and industry perception in this area (see Figure 18.2).

FIGURE 18.2
BP's new logo.

The company has come under heavy criticism for not practicing what it preaches, including the facts that although it has committed to invest heavily in solar and alternative energies, these investments pale in comparison to their investments, holdings, and revenues from nonrenewable and "dirty" energy. The company also has a poor safety record and other problems.

Still, the execution of their rebranding was brilliantly handled, including materials geared specifically for employees to understand the organization's new values and policies. This is an example of the issue we examined for SRI funds, where acknowledging and supporting the "best of breed" within an industry are at odds with the balance of claims and performance for the same company. To be sure, BP's stated strategic changes are much more positive, in sustainable terms, than those of their competitors. However, they still fall short of the ideal we want to see. Do we praise the company for the gains it does make or withhold that praise until it has reached sustainable perfection? Are its claims merely greenwashing or part of the process of corporate aspiration and change?

Clif Bar

One of the best examples of an organization that is reaching toward "restorative" status on the Sustainability Helix is Clif Bar, a San Francisco Bay Area company that specializes in sport and energy bars. Clif Bar was started in 1990 by Gary Erikson to produce great-tasting energy bars for his 175-mile-long bicycle rides. At the time, the only bars available were nutritionally impressive but tasted oppressive. In 1992, he started selling his bars, named after his father, at a local bakery and quickly found a following among bicyclists and climbers. By 2004, revenues were over $100M/year and Clif Bar had about 12 percent of the entire market for energy bars (see Figure 18.3).

FIGURE 18.3
The Clif Bar logo.

Clif Bar has always strived to express its values through their products, policies, brand, and actions. In 2003, organic ingredients became important to the company and steadily since, these have made their way into all products (70 percent of all ingredients), some of which (like the Nectar bar) are entirely organic. The company has also adopted sustainable principles in all of its operations, researching and developing new products, approaches, and policies toward these goals.

Clif Bar raises funds for charities and projects that fit its values, and a portion of each worker's hours can be used to serve social and environmental needs. Recent goals seek to fight global climate change and even restore habitats and communities. These efforts include planting over 13,000 trees, volunteering to create houses for those in need, offsetting carbon in transportation, building windmills, and using biodiesel for shipping, where possible.

Clif Bar (continued)

One thing that differentiates Clif Bar is its careful and "quiet" approach to declaring progress and making claims. It doesn't advertise, and it doesn't promote its sustainability agenda or progress in a sensational manner on its Web site or when it appears at events. Most of the sustainable messages travel via word-of-mouth or in news stories, although there is plenty of information available inside their Web site for those who look for it. This approach earns them deep credibility from customers who care, and it doesn't appear to those who don't care. It's a decidedly appropriate way of targeting sustainable progress to those who care enough about the issues. To those who don't, it's simply a great-tasting bar.

CHAPTER 19

Conclusion

A s stated in the introduction, design is both part of the problem and part of the solution to sustainability agendas. The history of design is inconsistent in terms of impacts, both positive and negative, along social, environmental, and financial criteria. Designers and developers have created amazing and wonderful things for the world and for people. Unfortunately, we've also created some solutions that have hurt people and the environment in a myriad of ways. However, this doesn't have to be the future of design.

We don't get to create meaning or change society very often, and we need to set realistic expectations about how long change takes and what part we can play in it. We're neither the only cultural actors nor the most influential. However, because we're involved in the process of creating new solutions, communications, and understandings—at significant points—we can have considerable influence if we choose to use it.

Designers need to decide which values they want to reinforce. There will always be a market for styles based on trends and fads. It's up to each of us, however, to decide to what extent we want to support these and how appropriate they are. The same goes for cultural messages we respond to and reinforce.

For example, I consider the term *retail therapy* to be one of the most dangerous concepts ever invented by marketers and reinforced by designers. The idea that people will feel better if they buy something new is sad and menacing. When we help people believe that they will be happy, or feel younger, or be more attractive by buying something, most of the time we are lying to them. There are exceptions, but these good expectations occur when we align emotions, values, and meanings with product and service attributes and trigger these effects legitimately.

We can choose to work for organizations with conflicting values from us (these are different for everyone: for some, these might be cigarette companies, for others, toy companies), or we can choose to work on projects that support, create, and sustain the world we envision. When we work against our own values, we dishonor ourselves. When we work to deceive our customers about themselves and their lives—and they often will never realize it—we do a grave disservice to them and to the world.

Designers need to decide which values they want to reinforce.

There are times when design does make us look and feel better about ourselves and the world— for example, a dress that accentuates a curve just so, a scent that reminds us of a treasured memory, or a poster that reminds us we're not alone in our outrage or admiration. Where our skills are employed with legitimate means, these are when we use our skills to their best—and most sustainable—potential. But there are other times when we carelessly (or carefully, even) use our skills to convince the insecure (such as teenagers) that they'll be sexy and attractive if they just buy this shampoo or that a new car will make them feel better in the morning. Emotions dissipate quickly and playing shallowly to them isn't sustainable. Customers feel dissatisfied and deceived when the things they were convinced they needed desperately or would make them feel special fall short and disappoint them. In addition, these are more things that didn't need to be produced in the first place since they performed no lasting purpose and gave no lasting benefit. This is the least sustainable kind of design of all.

I consider the term *retail therapy* to be one of the most dangerous concepts ever invented by marketers and reinforced by designers.

We never truly had the luxury to make (literally) tons of stuff, move it around the world, and sell some of it to people who didn't need it, the rest heading for a landfill. Today, these costs we never paid in the past are finally coming due—some quickly, others gradually. They are environmental, social, and market costs, and with an ever-increasing population that is ever more interconnected, these costs aren't hidden any longer. Design and related fields no longer have to be a contributor to these costs. We have the technology. We can make things better than they were before (in all aspects)—better, stronger, faster, and more sustaining of culture, people, profits, and the planet.

There is no design industry, really. The few organizations that connect designers are not official, nor powerful, nor controlling. Design has always been in the hands of designers, every one of us, and we have the power to, collectively, change the course of design. The frameworks and strategies described in this book are one set of tools to use to direct design in a more healthful path. They aren't the only tools, for sure. But, only we can choose to employ them in our personal and professional lives. There's no one standing over us judging us. There's no international body directing us how to design. No one is forcing us to reinvent or ignore our own values. We are the only ones who can make a change...

...and there are a lot of us.

Appendix A

Super Summary and Checklists

There's a lot to keep in mind when developing new products and services sustainably. In the beginning, it's difficult to consider every strategy or principle or balance every aspect of impact. So, as you gain more and more experience designing sustainably, here are two sets of checklists to use. The first is very reduced but helpful since it encompasses the most important concepts to consider. The second is more detailed and can be used further along in the development process.

Basic Checklist

- **Create More** (value, meaning, and performance) **with Less** (materials, energy, and virgin materials).

- **Focus on efficient and healthy alternatives.**

- **Use and promote local energy, resources, and labor wherever possible** (reducing transportation).

- **Don't use PVC at all** (or wherever there's an alternative).

- **Design solutions to be savored.**

Detailed Checklist

Preparation

- Learn about sustainability, take training courses, read books, attend conferences, and get certified if it will help (such as LEED certification if you're an architect).

- Get into the conversation about the sustainability in your area (or nationally). You need the community as much as it needs your participation and perspective.

- Teach your co-workers, clients, and partners about sustainability. At least, buy them a book or forward good articles to them. You'll have more success emphasizing efficiency, cost reductions, risk mitigation, and health improvements than improving ecological or social impacts.

- Start talking about what you're doing, but keep it understated for now. Consider a personal or corporate blog or RSS feed that your friends, families, co-workers, and clients can use to keep abreast of sustainability developments.

- Get sustainability principles into the mission, vision, and values of your organization. This may take some time, so get the conversation started now.

- Start measuring social and environmental impacts and benefits alongside financial impacts and benefits. As you measure and track these, you'll be better equipped to make integrated bottom line decisions across all three instead of merely on the basis of money.

- Join the Designer's Accord: **www.designersaccord.org**.

- Implement sustainability in your organization in small ways and grow these over time. Even small changes (like switching to recycled paper or mugs instead of disposable cups or banning water bottles) will help change behaviors and raise the level of comfort among your co-workers.

- Consider working with partners and even competitors to raise standards, share information, and even pool resources to promote the adoption of new standards or materials and lower costs by raising the volume of demand.

- Promote recycling and composting at home and at work—even if you have to take it to the recycling center yourself.

Development

- Start with manufacturing. The more impact you can make in the production of products and services, the lower its impact on the environment may be.

- Use eco-design strategies appropriate to the product or service solution.

- Reduce the overall material content and increase the percentage of recycled material in products.

- Reduce product and service energy consumption (of all types).

- Reduce the energy consumption in the manufacturing, recycling, transporting, and disposal phases as well.

- Reduce product and service water consumption.

- Eliminate toxic materials from product and service production and use.

- If toxic materials are unavoidable, make these easy to remove and separate them for recycling.

- Design more durable solutions that stay effective longer.

- Develop solutions that become precious in a way that people don't want to part with them.

- Eliminate unused or unnecessary product features.

- Design products to be quickly and easily disassembled.

- Consider transforming products into services by focusing on the value and benefits they provide to customers. Consider leasing and renting solutions in addition to those that rely on purchasing.

- Consider social issues as well as environmental ones. Who makes and services the solutions? Where? How? At what cost?

- Consider wider environmental issues, such as biodiversity, decentralization, competition, cooperation, and so on.

- Create and support "take-back" programs, either through retailers, distributors, or directly.

- Explore business models that support more sustainable solutions.

- Consider distributed manufacturing, servicing, and repair to localize economies and reduce transportation costs.

- Create upgradable, serviceable, and repairable solutions.

Materials

- Specify sustainably-grown materials when using paper, cloth, wood, or other organic materials.

- Choose materials based on recyclability, production waste, toxicity, weight, and reusability over renewability.

- Source materials, where possible, with the highest recycled and post-consumer recycled content.

- Work with suppliers and manufacturers to explore materials with better social and environmental performance, less toxicity, or better durability.

Promotion

- Calculate costs and impacts over the product or service's entire life cycle in order to realize the greatest impacts and benefits.

- Make claims that are transparent and verifiable. If you aren't going to "show your work" by making available calculations or details, don't bother making the claim.

- Use independent certification or ratings services where you can afford to do so.

- Become knowledgeable of government and third-party standards in order to make claims honestly and accurately.

- Test your claims with your friends or independent customers. If they don't easily understand what you've claimed, you're in the greenwashing zone.

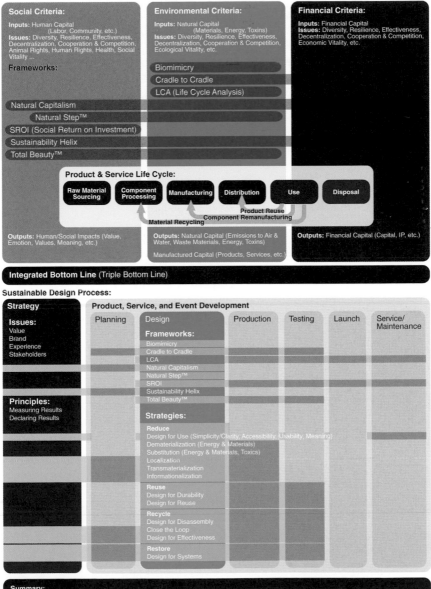

Social Criteria:

Inputs: Human Capital
(Labor, Community, etc.)
Issues: Diversity, Resilience, Effectiveness,
Decentralization, Cooperation & Competition,
Animal Rights, Human Rights, Health, Social
Vitality ...

Frameworks:

Natural Capitalism

Natural Step™

SROI (Social Return on Investment)

Sustainability Helix

Total Beauty™

Environmental Criteria:

Inputs: Natural Capital
(Materials, Energy, Toxins)
Issues: Diversity, Resilience, Effectiveness,
Decentralization, Cooperation & Competition,
Ecological Vitality, etc.

Biomimicry

Cradle to Cradle

LCA (Life Cycle Analysis)

Financial Criteria:

Inputs: Financial Capital
Issues: Diversity, Resilience, Effectiveness,
Decentralization, Cooperation & Competition,
Economic Vitality, etc.

Product & Service Life Cycle:

| Raw Material Sourcing | Component Processing | Manufacturing | Distribution | Use | Disposal |

Product Reuse
Component Remanufacturing
Material Recycling

Outputs: Human/Social Impacts (Value,
Emotion, Values, Meaning, etc.)

Outputs: Natural Capital (Emissions to Air &
Water, Waste Materials, Energy, Toxins)

Manufactured Capital (Products, Services, etc.)

Outputs: Financial Capital (Capital, IP, etc.)

Integrated Bottom Line (Triple Bottom Line)

Sustainable Design Process:

Strategy	Product, Service, and Event Development					
Issues: Value Brand Experience Stakeholders	Planning	Design	Production	Testing	Launch	Service/ Maintenance
		Frameworks: Biomimicry Cradle to Cradle LCA Natural Capitalism Natural Step™ SROI Sustainability Helix Total Beauty™				
Principles: Measuring Results Declaring Results		**Strategies:** **Reduce** Design for Use (Simplicity/Clarity, Accessibility, Usability, Meaning) Dematerialization (Energy & Materials) Substitution (Energy & Materials, Toxins) Localization Transmaterialization Informationalization **Reuse** Design for Durability Design for Reuse **Recycle** Design for Disassembly Close the Loop Design for Effectiveness **Restore** Design for Systems				

Summary:
More (Value, Experience, & Meaning) **for Less** (Materials & Energy), Emphasize Efficiency & Health, No PVC or Nuclear Power.

FIGURE A.1

This overview of the book's contents organizes the book's major principles, sustainability frameworks, and design strategies, indicating where in the development process each is most useful.

Appendix B

Resources

Books

Overview and Issues
Silent Spring, Rachel Carson

The Ecology of Commerce, Paul Hawken

The Joyless Economy: The Psychology of Human Satisfaction, Tibor Scitovsky, Oxford University Press

Frameworks
Natural Capitalism, Paul Hawken, Amory Lovins, Hunter Lovins

Cradle To Cradle, McDonough, Braungart

The Total Beauty of Sustainable Products, Edwin Datschefski

Biomimicry, Janine Benyus

The Natural Step for Business, Brian Nattrass, Mary Altomare

Design
Sustainable Fashion and Textiles, Kate Fletcher, Earthscan Publishers

Designer's Atlas of Sustainability, Ann Thorpe, Island Press

Design + Environment, Helen Lewis, John Gertakis

The Green Imperative, Victor Papanek

Design for the Real World, Victor Papanek

Sustainable Solutions, Martin Charter, Ursula Tischner

Green Design: Design for the Environment, Dorothy Mackenzie, 1991

The Complete Guide to Eco-Friendly Design, Poppy Evans, 1997

Trespassers: Inspirations for Eco-efficient Design, Ed Van Hinte, Conny Bakker, Ed van Hinte, 1999 010 Publishers

Sustainable Solutions: Developing Products and Services for the Future, edited by Ursula Tischner and Martin Charter

Organizational Change and Leadership
Leading Change Toward Sustainability, Bob Doppell

Mid-Course Correction, Ray Anderson

Beyond The Bottom Line: Putting Social Responsibility to Work for Your Business and the World, Joel Makower

The Next Sustainability Wave: Building Boardroom Buy-in, Bob Willar

The Anatomy of Change: A Way to Move through Life's Transitions, Richard Strozzi-Hecker

Organizational Change for Corporate Sustainability, Dexter Dunphy and Andrew Griffiths

Cannibals with Forks: The Triple Bottom Line of 21st Century Business, John Elkington

Eco-Economy: Building an Economy for the Earth, Lester R. Brown, 2001 W.W. Norton & Company

Sustainable Strategic Management: Strategic Management, W. Edward Stead, Jean Garner Stead, and Mark Starik

Articles
Development as if the World Matters, World Affairs Journal, 27 June 2005, **www.climatemanual.org**

Online Documents
Climate Protection Manual for Cities (Nat Cap Solutions): **www.idsa.org/whatsnew/sections/**

Guideline for Environmentally Responsible Packaging: **packaging.hp.com/enviro**

IDSA—Business Ecodesign Tools:
www.idsa.org/whatsnew/sections/ecosection/tools.htm

Ecospecifier.rmit.edu.au

Okala Teaching Guide:
www.idsa.org/whatsnew/sections/ecosection/okala.html

LASER Manual: **www.natcapsolutions.org/LASER.htm**

Eco-indicator 99 Manual for Designers:
www.pre.nl/download/EI99_Manual.pdf

Leverage Points: Places to Intervene in a System, Donella Meadows, The Sustainability Institute, 1999:
www.sustainer.org/pubs/>Leverage_Points.pdf

Evaluating the True Costs and Benefits of Bread, Murray Rudd:
www.sustainableventures.us/download/M_Rudd_SV_ODB_FINAL_082306.pdf

Web Sites

General Issues and Overviews
www.sustainabilitydictionary.com

www.saatchis.com/birthofblue

Frameworks
The Natural Step: **www.naturalstep.org**

Life Cycle Analysis Software: **www.pre.nl**

Life Cycle Analysis Tools: **www.gabi-software.com**

Life Cycle Analysis Tools: **www.ecoinvent.ch**

Life Cycle Analysis Tools: **www.simapro.com**

CMU Green Design Institute: EIO-LCA: **www.eiolca.net**

Biomimicry: www.biomimicry.net

Biomimicry Guild: www.biomimicryguild.com

Sustainability Helix: www.natcapsolutions.org/HELIX.htm

Pharos Building Materials Rating System: www.pharosproject.net

Global Reporting Initiative: www.globalreporting.org

Global Reporting Initiative system summarized by Covine:
www.covive.com/gri

Ratings Web Sites
www.alonovo.com

www.taoit.com

www.betterworldshopper.com

www.ethiscore.org

www.hrc.org/cei

www.c2ccertified.com

KLD Research: www.kld.com

Innovest: www.innovest.com

www.balancedscorecard.org

Sustainable Design
Environmentally Sustainable Product Design: www.espdesign.org

Demi—Overview on sustainable design: www.demi.org.uk

Locus Research—Sustainable product design: www.locusresearch.com

US Environmental Protection Agency—Design for Environment:
www.epa.gov/dfe

The Center for Sustainable Design: **www.cfsd.org.uk**

AIGA Center for Sustainable Design: **sustainability.aiga.org**

Designer's Accord: **www.designersaccord.org**

Sus Pro Net: **www.suspronet.org**

Blogs

www.worldchanging.com

www.treehugger.com

www.greenbiz.com

www.biothinking.com

www.triplepundit.com

www.ecotecure.com

www.02.org

Materials and Resources

Ecolect: **www.ecolect.net**

Rematerialise—Database of more sustainable materials: **www.rematerialise.org**

Transdesign Design materials Research: **www.transstudio.com**

American Chemistry's Plastics Learning Center: **plasticsresource.com/recycling**

US EPA—Green chemistry: **www.epa.gov/greenchemisry/index.htm**

APME—Life cycle analyses of plastics: **www.apme.org**

Ecospecifier: **www.ecospecifier.org**

Sustainable Packaging Coalition: **www.sustainablepackaging.org**

Materials Scraps: **materialscraps.com**

Global Recycling Network: **www.grn.com**

Pollution prevention World Information Network: **www.p2win.org**

Ingeo Fiber Library: **www.ingeofibers.com**

Business Resources

The Centre for Social and Environmental Accounting Research:
www.st-andrews.ac.uk/management/csear/

Sustainable Measures—How to measure sustainability:
www.sustainablemeasures.com

Online Magazine—Sustainability in Business:
www.greenfutures.org.uk

www.projectbetterplace.com

Educational Programs

MBA in Sustainable Management, Presidio School of Management:
www.presidiomba.org

MBA in Sustainable Enterprise, Dominican College:
www.greenmba.org

MSc in Holistic Science, Schumacher College:
www.schumachercollege.org.uk

MBA in Sustainable Business, Bainbridge Graduate Institute:
www.bgiedu.org

Index

L

M

ACKNOWLEDGMENTS

If it takes a village to raise a child, it surely takes a community to write a book. While I am listed as author, I couldn't have written this book without the support of many people.

To start with, I have to thank Lou Rosenfeld, Marta Justak, and Sue Honeywell who physically made this book possible. They're the main folks at Rosenfeld Media who signed, edited, and designed the book. In particular, Lou took a chance on a very different topic for one of his books that not all of his advisors thought fit well with the rest. It's a tribute to his commitment in not only creating a new kind of publishing company but also to his personal interest in sustainability.

I also need to express a deep gratitude to the entire community of faculty, staff, and students at Presidio School of Management, where I received my MBA in Sustainable Management in 2006. This is where, for two years (and three since), I learned about sustainability, studied examples, and built new ones. In particular, Hunter Lovins, Maggie Winslow, Paula Theilen, Nicola Acutt, Dwight Collins, Paul Sheldon, and Bob Dunham taught me more than I ever thought possible about our future. Although there are many students whom I've had the pleasure of working with and even teaching, my core study group, Holly Coleman, Meg Escobosa, and Ruth Katz got me through my two years with humor and a lot of great food. (We weren't self-named "Eating for Six" for nothing.)

My co-professors at the University of California, Berkeley, Sara Beckman and Alice Agogino, welcomed me with open arms into the course they have been teaching for over 13 years and helped me learn how to be a better teacher of design and sustainability. Together, their students and mine at CCA have created some wonderful solutions in the two years I've taught with them.

My friends and family also made this book possible. I've ignored them all too much while sequestering myself this year to get four books out at once. My mother, Phyllis Shedroff, brother Daniel and his family, friends, Laurie

Blavin, Nathalie Kakone, and their daughter Amelia have taken the brunt of this neglect. I also wanted to thank my friend Eric Friedman for making sure I didn't completely ignore fun and relaxation with trips to Sea Ranch or merely walks in the Marin Headlands.

Also, I appreciate the many people who allowed us to use their photos and diagrams to help explain how sustainability and design can change the future for the better.

I'm sure there are others, including the many authors I've read within this domain and my friends' mothers, Norma Laskin and Gail Solomon, who always appreciate being mentioned in my books even when they didn't know about them.

—Nathan Shedroff, January 20, 2009

PHOTO AND ILLUSTRATION CREDITS

Figure 1.1   Data Source: United Nations, World Populations Prospects, 2006 Revision

Figure 3.3   Photos courtesy of Ford Motor Company

Figure 3.5   Illustration courtesy of Biomimicry Guild

Figure 3.8   Photo courtesy of PAX Streamline: **www.paxstreamline.com**

Figure 3.13   Photo one courtesy of Toyota Motor Sales, USA, Inc.

Figure 3.13   Photo two courtesy of Robert Himler

Figure 3.20   Photo courtesy of Eva Solo

Figure 5.1   Photo two courtesy of Apple, Inc.

Figure 5.2   Photo courtesy of www.ifixit.com

Figure 5.3   Photo courtesy of Mark Pang

Figure 6.1   Photo courtesy of Herman Miller

Figure 8.2   Photo courtesy of CityCarShare

Figure 8.3   Photo courtesy of Interface FLOR

Figure 9.1   Photo courtesy of Architecture for Humanity

Figure 10.1   Photo courtesy of Artechnica

Figure 10.2   Photo courtesy of Better Place

Figure 10.3   Photo one courtesy of Apple, Inc.

                Photo two courtesy of Graham Skee

                Photo three courtesy of Flickr User TheDamnMushroom

                Photo four courtesy of Ellen Hobbs

                Photo five courtesy of Flickr user Legolam

Figure 10.4   Photo courtesy of Dyson, Inc.

Figure 11.1   Photos courtesy of Maille

Figure 11.2   Photo courtesy of Rapioli

Figure 13.3   Photo courtesy of William Good

Figure 14.1   Photo courtesy of Rickshaw Bags

Figure 15.1   Photo courtesy of Morio

Figure 15.2   Photos courtesy of Terracycle

Figure 15.3   Photo courtesy of Segway, Inc.

ABOUT THE AUTHOR

 Nathan Shedroff is the chair of the ground-breaking MBA in Design Strategy program at California College of the Arts (CCA) in San Francisco. This program melds the unique principles that design offers business strategy with a vision of the future of business as sustainable, meaningful, and truly innovative—as well as profitable.

Nathan is one of the pioneers in experience design, an approach to design that encompasses multiple senses and requirements and explores common characteristics in all media that make experiences successful—and has played an important role in the related fields of interaction design and information design. He is a serial entrepreneur, works in several media, and consults strategically for companies to build better, more meaningful experiences for their customers.

Nathan speaks and teaches internationally and has written extensively on design and business issues, including Experience Design 1.1, and Making Meaning, co-written with two members of Cheskin, a Silicon Valley-based strategy consultancy, which explores how companies can specifically create products and services to evoke meaning in their audiences and customers. Nathan is also the editor of the Dictionary of Sustainable Management, a Web site and now printed book. In addition, he maintains an extensive set of resources on experience design.

Nathan earned a BS in Industrial Design, with an emphasis on Automobile Design from the Art Center College of Design in Pasadena. However, fear of Detroit, coupled with a passion for information design, led Nathan to work with Richard Saul Wurman at The Understanding Business. Later, he co-founded vivid studios, a decade-old pioneering company in interactive media and one of the first Web services firms on the planet. vivid's hallmark was helping to establish and validate the field of information architecture by training an entire generation of designers in the newly emerging Web industry. Nathan was nominated for a Chrysler Innovation in Design Award in 1994 and 1999, and a National Design Award in 2001. In 2006, Nathan earned a Masters in Business Administration at Presidio School of Management in San Francisco, the only accredited MBA program in the U.S. specializing in Sustainable Business.

www.nathan.com

www.designmba.org

www.experiencedesignbooks.com